The International Library of Bioethics

Founding Editors

David C. Thomasma
David N. Weisstub
Thomasine Kimbrough Kushner

Volume 92

The *International Library of Bioethics* – formerly known as the International Library of Ethics, Law and the New Medicine comprises volumes with an international and interdisciplinary focus on foundational and applied issues in bioethics. With this renewal of a successful series we aim to meet the challenge of our time: how to direct biotechnology to human and other living things' ends, how to deal with changed values in the areas of religion, society, and culture, and how to formulate a new way of thinking, a new bioethics.

The *International Library of Bioethics* focuses on the role of bioethics against the background of increasing globalization and interdependency of the world's cultures and governments, with mutual influencing occurring throughout the world in all fields. The series will continue to focus on perennial issues of aging, mental health, preventive medicine, medical research issues, end of life, biolaw, and other areas of bioethics, whilst expanding into other current and future topics.

We welcome book proposals representing the broad interest of this series' interdisciplinary and international focus. We especially encourage proposals addressing aspects of changes in biological and medical research and clinical health care, health policy, medical and biotechnology, and other applied ethical areas involving living things, with an emphasis on those interventions and alterations that force us to re-examine foundational issues.

More information about this series at https://link.springer.com/bookseries/16538

Eike-Henner W. Kluge

The Right to Health Care: Ethical Considerations

Springer

Eike-Henner W. Kluge
Department of Philosophy
FRSC, University of Victoria
Victoria, BC, Canada

ISSN 2662-9186 ISSN 2662-9194 (electronic)
The International Library of Bioethics
ISBN 978-3-030-93837-6 ISBN 978-3-030-93838-3 (eBook)
https://doi.org/10.1007/978-3-030-93838-3

This Springer imprint is published by the registered company Springer Nature Switzerland AG
The registered company address is: Gewerbestrasse 11, 6330 Cham, Switzerland

For Heather

Preface

At first glance, the title of this study—*The Right to Health Care: Ethical Considerations*—indicates clearly and unambiguously what the study is all about: namely, that it deals with the question whether, ethically speaking, health care is a right, what ethical arguments can be adduced either for an affirmative or a negative answer, and what consequences each of these answers entails.

However, that clarity may be more deceptive than real, for it assumes that the notion of health itself is clear, that the concept of ethics is unambiguous, and that the notion of a right is not open to interpretation. However, there are at least four distinct ways in which the notion of a right may be interpreted: legally, socially, naturally, and ethically. Likewise, "health" may be understood in different ways: statistically, valuationally, and functionally being but three obvious examples; and the term "ethics" is also ambiguous: It may be understood from a philosophical, a religious, a social, and even a cultural perspective.

Therefore, an initial task of this study is to provide clarity about its subject matter, which will be done in the first chapter. More specifically, the first chapter deals with the notions of rights, ethics, and health, respectively and indicates how the relevant terms will be understood in the rest of the study and why they will be understood in that way. The following chapters will then present reasons why health care should be considered a right rather than a commodity, deal with the extent of that right and with what this entails for those who are engaged in providing health care in a professional or an institutional capacity. A final chapter will consider what all of this means for the individuals to whom the right belongs: namely, persons. Among other things, it will be suggested that there are certain obligations that belong to societies in which health care is provided but that, by the same token, this also imposes certain obligations on the members of the societies themselves. A special chapter is devoted to the ethics of resource allocation, both at the level of healthcare frameworks—i.e., at the level of policy or what is often referred to as macro-allocation—and at the hands-on level, or what is often referred to as micro-allocation.

There is considerable overlap between the topics that are discussed in some of the chapters, which means that there is a fair amount of cross-referencing between them. This is both inevitable and intentional. It is inevitable because essentially similar issues arise for the various players in the field of health care. It is intentional because although the issues are similar and sometimes even the same, they arise in different ways. This makes it important to discuss how this is the case and what way their respective implications differ—which can be done only by relating the various issue to each other.

Moreover, the different chapters of this study tend to be somewhat repetitious in the sense that essentially similar reasoning is used to support the particular claims that are being made. Again, this is both inevitable and intentional. It is inevitable because, in one way or another, the various claims are all ultimately grounded in the same ethical principles and, being interrelated, references to the same principles are bound to occur. It is intentional because by citing the same principles, it is possible to show clearly how the interrelationship between the claims comes about, and that the study as a whole presents an integrated argument for the right to health care and what this means.

The Appendix to this study expands the analysis somewhat by considering what implications the right to health care has for corporations that supply some of the tools that are used when providing health care, with particular focus on pharmaceutical corporations. The reason for separating this analysis from the core of the study is that while such corporations are engaged in the field of health care, they are not healthcare providers in the strict sense of that term and therefore are not subject to quite the same considerations that apply to society itself or to healthcare institutions and healthcare professionals. These corporations are commercial entities, and their mandate is not to provide health care but to provide their stockholders with a return for their investment. The fact that they do so by developing and supplying some of the tools that are used in healthcare delivery is merely their way of fulfilling their stockholder-directed obligations.

Care has been taken to support whatever points that are made either by indicating what factual support for such claims can be found or by citing appropriate authorities. This has the effect of making the study somewhat reference-heavy. However, this is a study in applied ethics. Applied ethics does not consist in developing theories or designing hypotheticals but in applying ethical principles to real life. Therefore, although hypothetical situations are sometimes constructed, this is only done to illustrate a point when it would be too lengthy to provide an actual example; and although theoretical considerations play an important role in the conduct of the analysis, they are used only to outline the ethical structure of the various situations.

An almost inevitable outcome of this study is that if the reasoning that it contains is sound, then it will challenge the validity of certain healthcare-related practices. That should not be unexpected. Not everything that happens in real life is ethical, policies and laws not being excluded, nor is tradition necessarily a sound guide to what is ethical.

Finally, this study deals with the right to health care, not with the right to health. This may strike the reader as peculiar, particularly since the United Nations Committee on Economic, Social and Cultural Rights, in its "General Comment No. 14", states that there is a fundamental right to "the highest attainable standard of health",[1] and in turn bases this on Article 25 of *The Universal Declaration of Human Rights.*

However, as a closer reading of the comment shows, the right in question is not the right to health but the right to the highest *attainable* standard of health. The two are not the same thing. Moreover, what counts as the highest attainable standard of health is functionally related not only to the prevailing conditions in a given society but also to the resources that the society has available for providing health care; because, as will also be discussed later, there are other types of services that a society is ethically obligated to provide for its members, and each of these types of services is resource-intensive. Consequently, a given society can apportion only a limited amount of its resources to providing health care, and there will be variations in the levels and types of healthcare services that the individuals in different societies can claim as a matter of right. This in turn means that while a so-called developed nation is obligated to provide its members with different—and indeed more—healthcare services and at a higher level than a so-called underdeveloped nation, this is not inherently unethical. The resource pools that are available to them respectively are simply different. However, it would be unethical if the various societies or nations did not provide healthcare services at all, or if they did not adhere to the principles and conditions that will be outlined in the chapters that follow.

Victoria, Canada Eike-Henner W. Kluge

[1] https://undocs.org/E/C.12/2000/4.

Contents

Chapter 1
A Definitional Preamble

Abstract This chapter introduces and defines the term that will be used in the rest of the study, and to indicate what role they play in the analysis of the right to health care. It also distinguishes between different approaches to analysing the right to health care and indicates why an ethical approach will be used in the subsequent analysis.

Keywords Ethics · Health · Person · Right

1.1 Introduction

At first glance it seems quite clear from the title—"The Right to Health Care: Ethical Considerations"—what this study is all about: namely, the ethics of health care when health care is construed as a right. A moment's reflection, however, suggests that this clarity may in fact be more deceptive than real, for the term "health" can be interpreted in various ways, as can the term "ethics", to say nothing of the term "right." The aim of this chapter, therefore, is to remove any such ambiguity and to indicate precisely how these terms will be understood in the subsequent analysis and why they will be understood in this particular way.

1.2 Right

To begin with the term "right:" there are at least four different ways in which the term when taken as a noun can be understood: As an entitlement that is legally bestowed on a particular party, where the latter may be either singular or plural in number; as an entitlement that someone has to something on the basis of social convention; as an entitlement that belongs to individuals by their very nature; and as an entitlement that belongs to someone on the basis of ethical considerations.

E.-H. W. Kluge, *The Right to Health Care: Ethical Considerations*, The International Library of Bioethics 92, https://doi.org/10.1007/978-3-030-93838-3_1

1.2.1 Some Common Factors

These interpretations are, of course quite distinct both in their meanings and implications. However, this notwithstanding, all four interpretations have a common conceptual core: namely, the notion of entitlement. If someone has a right, then they have an entitlement, which in turn means that someone else has a duty to respect or even to act on that entitlement if that entitlement is claimed. A right, therefore, imposes a duty on someone. In that sense, "right" and "duty" are correlative terms. And just as the party to whom a right belongs may be either a singular or a collective entity, so the party who has the corresponding duty may be a singular or collective entity. This becomes important later on when dealing with health care considered as an ethical right. To look ahead, the right to health care, while it belongs to individual persons, imposes a duty both on individuals such as health care professionals as well as on the collectivity or society in which the individuals are embedded.[1]

The second thing that all four interpretations have in common is that if a right is an entitlement, then a right is something that the right-holder may choose or not choose to exercise. By contrast, a duty is something that does not involve choice. If and when the circumstances under which a right is claimed are operative then, all other things being equal, the party who has the duty in question must do their best to fulfil the corresponding duty on pain of being in violation of that duty. The reason for the limiting clause "all other things being equall" is that, as is generally agreed, one cannot have a duty to do what in fact is impossible under the circumstances that obtain.[2] The reason for this is not simply legal—the legal recognition of this goes back to ancient Rome, where it was expressed as "*ultra posse nemo obligatur*"[3]—but also logical. While it is possible to state an impossible task in a grammatically unexceptional fashion, logically speaking it involves a contradiction and hence is conceptually meaningless.[4]

A third factor that is important and that again all four interpretations have in common is that the exercise of a right involves values. More specifically, whether a right is exercised—i.e., whether an entitlement is claimed—is matter of decision on part those to whom the right belongs. Thus, even though rights are entitlements, this merely conditions the framework within which the right-holder is situated by structuring it in such a way that if the right-holder chooses to claim that entitlement

[1] See Chap. 3, *infra*.

[2] Where it is understood that the impossibility has not been brought about by the party whose duty it is to fulfil that right. See Chap. 2, pp. 23 f.

[3] Justinian, *Corpus Juris Civilis*; Celsus, *Digest of Roman Law* 8: "Impossibilium nulla obligatio est" (There is no obligation to do anything that is impossible) [D.50.17.185]. See also *Taylor v. Caldwell* [1863] EWHC QB J1.

[4] As will become important later when dealing with allocation of resources, impossibility is not the same as futility. Futility looks at whether the relevant action is likely to be successful in producing the intended or expected result. Thus, it would be futile to perform a colectomy on someone dying of inoperable metastasized cancer of the colon that has spread throughout the whole body and neither chemotherapy nor radiation have proved effective. It would be futile because it would not change the fact of impending death.

then—again, all other things being equal—the party who has the corresponding obligation should act in the appropriate manner. However, as was said before, the right-holder need not exercise their right. Whether they choose to do so is a function of their values. In that sense, values are action potentials[5] and play a crucial role in the functioning of rights. This becomes important when dealing with the allocation of resources. Thus, someone may have the right to a particular type of health service and, on occasion, this right may have priority over someone else's right to that same service. However, the right-holder may decline to exercise their right because of the values that they hold.

By contrast, volition plays no role in the context of duties, and values have only very limited relevance. That is to say, whether someone who has a duty wishes to fulfil that duty does not affect whether the duty in fact obtains. Once the enabling condition of a right is met and the right-holder claims that right, it is irrelevant whether the party who has the corresponding duty wishes to fulfil that duty. Failure to do so would constitute a default. The only exception would be if the values of the party who has the duty would be violated by the fulfilment of that duty if these values are values that are subscribed to by both by the right-holder and the duty-holder. More specifically still, the only exception would be if fulfilment of the duty would violate the values of the framework in which both the right-holder and the duty-holder are situated.

This brings the discussion to the different interpretations of the notion of a right that was indicated above: namely legal, social, natural and ethical interpretations.

1.2.2 Legal Rights

Legal rights are rights that are promulgated by a duly constituted authority and that belong to the individuals who are situated within the relevant juridical framework. Since juridical frameworks differ both in structure and content, the fact that something is a right in one jurisdiction does not entail that it is a right in another jurisdiction unless there are provisions or accords between the jurisdictions which stipulate that the provisions in question are legally recognized in both jurisdictions.

Further, while legal rights may recognize what has already been considered a social, natural or ethical right, these do not necessarily overlap. In other words, legal right are not necessarily coextensive with socially, naturally or ethically grounded rights, and may exceed or even fall short of those that are otherwise grounded.

A third point that distinguishes legal rights from other kinds of rights—at least in some jurisdictions—is that while legal rights are rights that are promulgated by a duly constituted authority, their existence as such only establishes them in principle. What a legal right actually amounts to in real terms may be a matter of interpretation

[5] Hodgkinson (1996).

by a duly sanctioned interpretative agency such as a court of law of the relevant jurisdiction.[6]

Finally—and this they have in common with socially grounded rights—since legal rights exist only within a societal context, they are also temporally bound to the existence and the state of that social context. Hence they do not hold for the relevant people for at all times, since the juridical framework that characterizes the particular society may undergo a radical shift.

1.2.3 *Social Rights*

Social rights are rights that are grounded in what may be called the social contract that binds the society in which they are recognized.[7] That contract—if indeed that is the correct word—is an agreement that is entered into either explicitly or tacitly by the members of society simply by being members of that society. It binds the members to accept certain rights both in their own behalf as well as on the part of other members of that society even if these rights are not explicitly recognized in law. In that sense—and abstracting from the fact that they obtain independently of legal recognition—they are limited in ambit to the society in which they are recognized.

It should be clear from this that, just like legal rights, social rights are limited in their domain to their specific social context. Therefore, since there is no actual unified, integrated and functioning global society—the United Nations do not count in this regard, since they are merely distinct societies that are associated for a particular set of purposes without thereby giving up their distinct natures—social rights do not necessarily have global validity. Furthermore, like legal rights, they are temporally bounded because they obtain only as long the relevant society expresses the characteristics that the society had when the social rights originally arose. It should also be clear from this that legal rights may reasonably be construed as a particular species of social rights.

1.2.4 *Natural Rights*

Natural rights are rights that are grounded in what is sometimes called natural law, which latter is the framework of laws that is independent of any human conventions that afford them validity. They belong to human beings simply because they are human beings. Moreover, unlike social and legal rights, they are independent of any social context and are presumed to have atemporal and global validity.[8] As soon as entities naturally satisfy a particular set of conditions—for instance being

[6] Sullivan (2003).

[7] See Gough (1936), Riley (1982), Rousseau (1762).

[8] Dworkin (1986), Finnis (1980), Fuller (1964).

human or, more generally, having the capacity for cognitive awareness and volition—these entities are said to have natural rights. By that token, however, all conceptions of natural rights ultimately have a philosophical core, which in turn means that, logically speaking, natural rights are merely a species of ethical rights. At the same time, it should be noted that the notion of natural rights is sometimes associated with a religious perspective such as that of Christianity, Islam or Judaism.[9] However, as is clear from the writings of Aristotle[10] and later philosophical and political thinkers,[11] that need not be the case.

If, then, natural rights essentially are a species of ethical rights, their validity ultimately reduces to that of ethical rights—which brings the discussion to the latter.

1.2.5 *Ethical Rights*

Ethical rights are justified claims that derive from the fundamental principles of ethics, where these latter are assumed to be objective in nature and to determine a framework within which all entities that satisfy certain criteria automatically enjoy these rights. Like natural rights, therefore, ethical rights are thought to obtain independently of legal or social conventions, and to hold globally as long as the criteria that engage them are met.

The immediate question that arises, therefore, is what these fundamental principles are—if indeed there are such principles—and whether they have objective validity.

1.3 Ethics and Ethical Approaches

This brings the discussion to the nature of ethics. The problem here is that just as is the case with the notion of rights, there are various conceptions of the nature of ethics. For instance, ethics has been defined as the "study of morality or moral conduct,"[12] the study of "moral principles that govern a person's behaviour or the conducting of an activity,"[13] the field of study that "involves systematizing, defending, and recommending concepts of right and wrong behavior,"[14] "a branch of philosophy that involves systematizing, defending, and recommending concepts of right and wrong

[9] Emon et al. (2015).

[10] Cf. Rommen (1947).

[11] Haakonssen (1996).

[12] *Encyclopedia Britannica*, available at https://www.britannica.com/demystified/whats-the-difference-between-morality-and-ethics.

[13] *Oxford English Dictionary*, available at https://en.oxforddictionaries.com/definition/ethics.

[14] *Internet Encyclopedia of Philosophy*, available at http://www.iep.utm.edu/ethics/.

conduct,"[15] "the science of the ideal human character and behaviour," and the study of "how one should live"[16]—to mention but a few.

What they all have in common, however, is that in one way or another they all characterize ethics as being concerned with right and wrong conduct. But here again there are several possible ways of interpreting this, because there are several ways in which standards of right and wrong conduct can be understood: most prevalently on the basis of religion, on the basis of social consensus, on the basis of tradition, and on the basis of purely philosophical considerations.

However, not all of these provide a globally valid framework, which is crucial for achieving the aim of the present study. For the sake of conceptual clarity, therefore, it may be appropriate to consider these variations in turn.

1.3.1 Religiously Based Ethics

The reason why religiously based ethics is of interest in light of the title of this study is that religiously based ethics has not only played an important historical role in the evolution of health care delivery both on a national as well as a global scale, but that it is still foundational to the care that is sometimes delivered in denominational hospitals, and also structures health care decision-making at the individual level when health care professionals and patients make ethical decisions based on their religious beliefs. Further, it grounds the present-day policies of some non-governmental agencies like the Aga Khan Health Services[17] and the Catholic Church, both of which provide health care on a global level.

However, one of the major problems with using religiously-based ethics to analyse the right to health care is that there are and have been quite literally hundreds of religions, and not all of them share the same principles or agree on what constitutes right and wrong conduct. Thus, the principles that underlay Mesoamerican religions with their emphasis on human sacrifice were fundamentally distinct from and incompatible with the principles that underlie Judaism, Islam or Shintoism, and the basic principles of the latter do not necessarily agree with each other or with those of Haida religion, Azande religion or Sikhism.

Moreover, as was already pointed out when considering the question whether rights have a religious basis, a religiously-based ethics, whatever its nature, depends for its probative force on belief. That limits the validity of a religiously based ethics to the framework of that belief—which in turn means that if the issue of the right to health care is to be resolved in a global and above all in an objective manner, a religiously grounded contender for assessing whether there is an ethical right to health care (and what form it should take) would have to establish that its belief system is objectively valid. However, no religiously based ethical framework has

[15] *Internet Encyclopedia of Philosophy*: Ethics. http://www.iep.utm.edu/ethics/.

[16] Williams (1985).

[17] Aga Khan Health Services, https://www.akdn.org/our-agencies/aga-khan-health-services.

been able to do that. This is not to deny that religiously based ethical theories do not make claims to the universal validity of their principles, or that religiously grounded deliverers of health care do not engage in what otherwise would be called meritorious work. However, making a claim is not the same as establishing the validity of that claim, and doing meritorious work does not establish that the belief-system that leads to such work is objectively valid. Therefore basing the analysis of the right to health care on an ethics that is religiously grounded would be valid only within the framework of that religion—which brings the issue back to the question of global and objective validity.

In other words, religiously grounded principles have binding validity only within the conceptual framework of that particular religion. Therefore unless it could be shown that the principles of all religions are merely different expressions of one and the same principles, a religiously grounded notion of ethics would not yield a globally valid interpretation of ethics.

1.3.2 Social Consensus Based Ethics

The social consensus approach to ethics is also incapable of providing a globally valid framework of ethics, and this for the simple reason that it can identify ethical principles only relative to the particular social setting in which the consensus exists. And even here it would face the same problem as a religiously based approach because there are many distinct societies in which there exists a distinct social consensus on what is right and what is wrong.

In fact, it would not even yield objectively valid ethical principles within the same social setting without temporal limitations—a point that was expressed in a somewhat roundabout fashion by Abba Eban when he reportedly said that "consensus means that everyone agrees to say collectively what no one believes individually."[18] The point that Eban was making was that social consensus is subject to the psychodynamics of a particular social grouping at a particular point in time. Since these psychodynamics shift over time, this means that a social consensus based ethics would only be binding for that group during the time in which the psychodynamics that resulted in such a consensus remained stable. Consequently, since there is no single unified global society and there is no universal social consensus that is psychodynamically stable, this approach cannot yield an objectively and globally valid framework for ethics either.

[18] Eban, Abba. "Quotes;" retrieved 25/04/2021 at https://www.brainyquote.com/quotes/quotes/a/abbaeban167934.html.

1.3.3 Legally Based Ethics

Understanding ethics in terms of the laws, legal standards or regulations that have been established by a duly empowered legislative or juridical body is also not an acceptable option for determining whether there is an ethical right to health care and what shape that right might take. The fact that a particular set of principles has been identified and promulgated by a duly empowered legislative or juridical body does not necessarily mean that these principles are ethically defensible. To assume that they are is to confound procedural correctness with ethical validity. Thus, the principles that underlay and structured the apartheid laws in the Union of South Africa in the 1950s were procedurally valid in that they had been enacted by a duly constituted authority in accordance with the relevant protocols. That, however, did not make then ethically defensible. A similar consideration applies to some of the laws that were enacted by Nazi Germany.

Moreover, such an approach to ethics also runs into the problem of relativity. If legally grounded ethical principles were to have global validity and not be relative to a specific legal system or a specific juridical tradition, there would have to be a universally binding set of legal and juridical principles. There is no such consistent and globally valid set of legal principles. In other words, there is no framework or tradition of legal principles to which all countries and all societies subscribe. The difference between a common-law as opposed to a civil-law approach illustrates this only too well. Therefore absent such a framework, the principles of legal ethics would again be framework-relative, and hence would fall subject to the same critique of relativism as all religiously or consensus based approaches. Hence, it would be a mistake to base the analysis of the right to health care on legal ethics.

1.3.4 Tradition Based Ethics

With due alteration of detail, similar considerations rule out tradition-based ethics as structuring a globally valid analysis of the right to health care. To put it bluntly, tradition only guarantees that whatever is enshrined in tradition or has a traditional basis is historically validated. However, being historically grounded or validated is not the same as being ethically grounded or validated. In fact, what is historically grounded or validated may be quite unethical. So, for instance, the practice of female genital mutilation is firmly grounded in the traditions of some cultures in Africa, Asia and the Middle East.[19] However, as the World Health Organization has pointed out, it is "a violation of the human rights of girls and women."[20] Such a claim would be nonsense if ethics was grounded in tradition. Likewise, discrimination on the basis of sex, gender and even skin colour is firmly rooted in the tradition of many societies

[19] Althaus (1997).

[20] World Health Organization (2017).

and cultures. However, as may be gathered from the *Universal Declaration of Human Rights*,[21] this is not ethically acceptable.

1.3.5 Philosophically Based Ethics

If the preceding considerations are cogent, then a philosophically oriented approach to ethics is the most likely contender for successfully identifying universal and objectively valid ethical principles and investigating the ethics of the right to health care on that basis.

However, even assuming that this is correct, it raises the question of which approach to philosophically based ethics is valid for, as in the case of the other approaches to ethics that have just been considered, there are various and distinct philosophical interpretations of ethics itself. Thus, deontological ethics differs significantly from utilitarian ethics, as do some versions of feminist ethics and virtue ethics.

Nevertheless—at least insofar as health care is concerned—despite the fact that these philosophical approaches are distinct in orientation, they all have something in common: Namely, they construe ethics as a framework of right and wrong conduct towards persons, and they generally interpret the notion of a person itself as an entity that has the capacity for cognitive awareness and volition.[22] Therefore it may be appropriate to consider this common notion before proceeding to take a closer look at these distinct approaches to philosophical ethics and evaluating their suitability for analysing whether there is a right to health care.

1.3.5.1 Persons

The first important point to note in this connection is that the above notion of personhood says nothing about the degree or acuity of cognitive awareness of an alleged person; nor does it say whether that the capacity must be actualized. Even feminist ethics, which tends to see personhood as functionally related to social embedding,[23] agrees that persons do not cease to be persons when they are not actually cognitively aware or have little if any sophisticated understanding. This point becomes important later on when dealing with the allocation of resources and with such issues as of the rights of unconscious individuals, small children or of individuals who are cognitively severely disabled or impaired, such as Alzheimer patients or those suffering

---

[21] United Nations, *Universal Declaration of Human Rights*, Article 1, https://www.un.org/en/universal-declaration-human-rights/.

[22] Environmental ethics and similar domains are not relevant to the present discussion because they are not involved in health care ethics insofar as it is the present enquiry is concerned.

[23] Cf. Sherwin (1991).

from dementia or trisomy 21. It also becomes important when it comes to substitute decision-making in the realm of health care.

The second important point is that there are different kinds of persons because there are different kinds of entities that have the capacity for cognitive awareness and volition. The standard case of a person, of course, is that of a living member of the species *Homo sapiens*: in other words, the case of a natural person in the traditional sense of that term.[24] However, groups of individuals may function in a unified fashion as an individual and as such may be aware of certain facts, have certain values and aims, and may make decisions on that basis. This is formally recognized socially, legally as well as ethically in the concept of a corporate or constructive person.[25] Like human persons, corporate or constructive persons have rights and duties. That it is why their actions can legitimately be evaluated using the same ethical principles as those that apply to biologically human persons, and why they can be held both held ethically and legally accountable—a point that will be discussed more fully later.[26] Further, social groupings such as indigenous societies may count as persons because, again, as a unit they may be aware of certain facts, have certain values and make decisions on that basis. This becomes important when it comes to such things as privacy issues in the domain of health care or to health care oriented research,[27] genetics and the like, as well as in the context of substitute decision-making.

1.3.6 Competing Philosophical Theories of Ethics

At the same time, the differences between these philosophically based ethical theories can be profound, and they can lead to different positions on how the issue of the right to health care is to be resolved, and indeed who has a right to health care.

1.3.6.1 Feminist Ethics

Like religiously based ethics, feminist ethics comes in various forms, ranging from theories which contend that the ethical framework that holds for women is different from the one that holds for men[28] to theories which hold that it is the difference in their respective social embedding that leads to a fundamental difference in ethical considerations between women and men,[29] and that in many ways women have a

---

[24] Teichman (1985).

[25] Cf. French (1979), Hess (2014), List and Pettit (2011).

[26] See Chap. 7, *infra*.

[27] For formal recognition of this at the government level, see Government of Canada, Research Involving the First Nations, Inuit and Métis Peoples of Canada; available at http://www.pre.ethics.gc.ca/eng/policy-politique/initiatives/tcps2-eptc2/chapter9-chapitre9/. See Wilson (2001).

[28] *Cf*. Firestone (1970).

[29] Gilligan (1982).

more sensitive ethical perspective than the one that is to be found in traditional philosophical ethics, which tends to be principle-oriented and male-based in origin.[30]

These differences notwithstanding, feminist ethical theories tend to have as their focus the fact that, for various reasons, traditional approaches to ethics have been disposed to minimize or even to dismiss the historical differences between female and male roles in human societies—a fact that allegedly has resulted in a skewed picture of the ethical status of women as persons. This appears to underlie at least part the reasoning that is involved in many feminist ethical discussions, particularly in discussions of reproductive ethics—a topic that has assumed increasing importance in health care ethics in recent years. Feminist ethical analyses have also been influential in shaping how the notion of care itself is to be construed[31] both from a theoretical as well as an applied perspective, and in the reasoning that surrounds the ethics of contemporary health care.[32]

At the same time, and notwithstanding their critique of the traditional principled approach to ethics, the logical focus of feminist ethical theories tends to be the individual as person, for this is ultimately what underlies the claim that women as persons should be recognized as having the same rights and duties as all other individuals precisely because they are persons. In that sense, irrespective of whether they are officially recognized as fundamental principles, equality, justice and autonomy play a logically crucial role in feminist ethics. In that regard, feminist ethical theories are logically indistinguishable from deontological ethical theories, and essentially differ from these only in their focus or perspective[33] and in the addition of parameters like empathy interdependence and community as well as in the role of emotions and feelings and the de-emphasis of hierarchical considerations.[34] Therefore, when all is said and done, feminist ethics is not really a rejection of deontological ethics but an expansion of the realm of considerations that are deemed relevant in deontologically based ethical deliberations.

In other words, rather than providing a distinct ethical framework for analysing the right to health care, feminist ethical theories differ from deontological approaches more in their emphasis than in their logic or in the principles to which they are ultimately committed. Where this is not the case, and in the absence of adherence to such principles, feminist ethics theories provide no effective mechanism for resolving conflicting claims to health care, whether that be at the hands-on or at the policy level.

That is to say, given that health care is an applied matter, and given further that no matter how the right to health care is ultimately understood—and even whether there is such a right—the resources that are available for providing health care are inherently limited. This means that discriminatory allocation decisions have to be made, whether that be at the policy level or at the level of hands-on health care. Consequently a conceptual framework that does not provide for a conflict resolution

---

[30] Lindemann (2005).

[31] Held (2006).

[32] Tong (2006).

[33] Baehr (2004), Hay (2013), Nussbaum (1999), Stone-Mediatore (2004).

[34] Jaggar and Bordo (1992).

mechanism will not deliver an applicable analysis of the ethics of health care, let alone of whether there is a right to health care and what form that right might take. Purely feminist ethical theories that focus on gender or sex based differences, therefore, are of limited usefulness when analysing the right to health care. It is only when such theories return to their deontological roots that they can provide such an analysis—in which case it is their deontological basis that does the analytical work, not their purely feminist perspective. It is for this reason that feminist ethics will not be used to investigate the ethics of the right to health care

1.3.6.2 Virtue Ethics

A philosophical approach that differs fundamentally in orientation from that of feminist ethics but to some degree overlaps with that of religiously based ethical theories is that of the group of ethical theories that is commonly referred to as "virtue ethics."[35] They include Platonist virtue ethics,[36] exemplarist virtue ethics,[37] eudaemonistic virtue ethics[38] and even agent based virtue ethics.[39] What they all have in common, however, is their contention that what constitutes rights and obligations—and this includes the right to health care[40]—should not be determined on the basis of so-called fundamental ethical principles but on the basis of whether the relevant actions or provisions are in keeping with an identified set of virtues.

However, using a virtue-ethics approach to analyse whether there is a right to health care and what form it might take is also not an acceptable option. The reason is that, just as is the case with religiously based ethical approaches, virtue-based ethical approaches are framework-dependent. More specifically, a virtue-theory analysis is valid only within the framework of the particular virtue-based ethical theory in question and of the virtues that it identifies as fundamental. However, as was pointed out a moment ago, there are several such contenders. Consequently, if a virtue-ethics based approach were to be used for dealing with the issue of the right to health care, then such analysis would be objectively and globally compelling only if there was a universally agreed-on set of virtues. And there is no such set.

Moreover, as is the case with pure feminist ethical theories, virtue-ethical theories lack an objectively valid way of resolving conflicts between competing claims to health care. In other words, just like pure feminist ethical theories, they do not provide an effective and objective conflict resolution mechanism when competing claims to health care cannot be satisfied together and a discriminatory decision has to be made. However, offering such a decision-making device is essential for any framework that purports to provide an analysis of the right to health care since, as was

[35] Dillon (2020).

[36] Chappell (2014).

[37] Zagzebski (2010).

[38] Badhwar (2014).

[39] Slote (2001).

[40] Dawson (2009), Drane (1995), Pellegrino (1993), Toon (2014).

previously pointed out, health care is a resource-intensive enterprise and resources are always limited.

Consequently virtue-based ethics cannot form the basis of an objective and globally valid analysis of the right to health. This is not to say that such an analysis—whatever it may ultimately turn out to be—may not identify certain policies or principles as being virtuous or as involving virtues. However, if that occurs then it will not be because such an analysis will be based on a particular virtue or set of virtues. Rather, it will result from the fact—if it is a fact—that it is the principles that structure the ethical framework of the analysis that will validate the claim that the respective policies or actions are virtuous. In other words, the attribute of being virtuous will be derivative, not basic.

1.3.6.3 Utilitarian Ethics

Utilitarian ethics is a teleological, outcome oriented or consequentialist approach to ethics. In broad outline, such an approach evaluates the ethical acceptability of a given action or policy on the basis of its outcome. More specifically, utilitarian theories tend to subscribe to what is sometimes called the Principle of Utility, which is to the effect that an action or policy is ethically acceptable if and only it results or is likely to result in the greatest amount of good and the least amount harm for the greatest number of people. Differently, and putting it in an imperative form, they stipulate that one should always act in such a way as to maximize the balance of good over harm for the greatest number of people.

The Principle of Utility itself leaves undefined what counts as good or harm—which in turn opens the door to different versions of utilitarianism. One such version identifies the good as pleasure,[41] another as happiness,[42] a third says that it consists in the attainment of certain ideals[43] and a fourth claims that it really consists in a combination of several of these factors. They are called hedonistic, eudaemonistic, ideal and mixed utilitarianism respectively.[44]

And just as there are differences within utilitarianism about what counts as the good, so there are differences about how the good is to be identified. One version claims that a special sort of insight or intuition is involved,[45] whereas another looks at social preference[46]—which of course means that this may change over time as social attitudes and preferences change. Further, utilitarian theories differ on whether moral predicates identify ethical qualities of actions or policies in a bivalent manner or on a scalar basis.[47]

---

[41] Bentham (2009), Mill (1998).

[42] Sidgwick (1981), Parfit (1986), Singer (2011).

[43] Moore (1912).

[44] Brandt (1979).

[45] Audi (2004).

[46] Hare (1981).

[47] Sinhababu (2018).

Finally, utilitarian theories may differ on how the test of utility is to be applied. Act utilitarianism proceeds on a case-by-case basis.[48] It evaluates the question whether there is a right to health care not on the basis of policy but by assessing each situation on its own terms, without reference to universal rules or guidelines. By contrast, rule utilitarianism maintains that utility is not something that can be calculated for individual cases but only for types of cases.[49] Therefore, on this approach, whether there is a right to health care and what this means in actual terms should be assessed at the policy level and the particularities of a specific situation would be irrelevant.

At first glance utilitarian ethics, in one form or another, offers an ideal basis for an objective and globally valid analysis of the right to health care. Thus, it does not discriminate on the basis of beliefs, traditions or social consensus, all of which are culture-bound and relative in nature. Moreover, it looks at persons insofar as they are socially embedded and functionally interrelated in real life, not as isolated entities whose claim should be evaluated on an abstract basis. Therefore it can provide an analysis of competing claims to health care as well as of competing duties by looking at what is the case in the actual social setting—something that is crucial when dealing with the issue of resource allocation. Third, it holds out the promise of being objective because it focuses on facts. It also appears to satisfy the criterion of global validity because it is independent of legal considerations. Finally, it seems to be perfectly compatible with the distinct notions of personhood that were identified above because it can be used to evaluate the actions of constructive as well as natural persons irrespective of their sex or gender.

However, these advantages notwithstanding, the adoption of a utilitarian perspective when analysing the right to health care faces several problems. One of the major issues that it faces is precisely on the notion of personhood that was just mentioned: To put it in a nutshell, it essentially divorces the notion of ethics from that of personhood.

More specifically, precisely because they are consequentialist in nature, utilitarian approaches evaluate the ethical acceptability of a policy or action solely on the basis of whether—and how—their outcome contributes to whatever is identified as "the good," however the latter may ultimately be defined. Consequently, a utilitarian perspective structures its framework of rights and duties not on the basis of how these derive from the inherent nature of persons as entities with the capacity for cognitive awareness and volition, but on the basis of the outcome of their actions: Differently, its focus is the individual as producer, not as person.[50]

In other words, purely logically speaking and at its most basic level, irrespective of the fact that utilitarian theories have always been subsumed under the headings of "ethics", utilitarian approaches differ from what traditionally have been called ethical theories by essentially shifting the framework of evaluation away from the notion of a person as an entity that has the capacity for cognitive awareness and volition and then evaluating the ethical aspects of their actions on the basis of how they use that capacity to the notion of an entity whose actions count solely as bringing about a

---

[48] Eggleston (2014).

[49] Rawls (1971).

[50] Mandal et al. (2016).

certain outcome. In other words, as a production unit. While utilitarian theories may agree that such units may have the capacity for cognitive awareness and volition, that is not what grounds their ethical evaluation of what they do or refrain from doing, or of their duties and rights. In other words, utilitarian approaches to ethics essentially confuse ethics with economics.

This last may require some further explanation. To put it first in general terms, while the number of persons who claim a right may determine whether their claim can be met, it does not determine whether the claim itself is valid. More specifically to the present context, while the number of persons who claim a right to health care may determine whether the available resources are sufficient to satisfy their claims, that does not determine whether their claims are legitimate, nor do the quantitative consequences of meeting their claims determine whether there is a right to health care in the first place. That is while measures such as quality- or disability-adjusted life years may be used to determine the comparative or downstream costs of fulfilling a claim to health care[51] and even of how a health care allocation policy frameworks should be structured from an economic perspective, logically speaking these cannot be used to analyse the question of whether there is right to health care either in general or in a given case.[52]

Moreover, there is an important consideration that applies in this connection. As was just pointed out, by their very nature utilitarian analyses of the right to health care can only provide an answer to the question whether there is a right to health care on the basis of outcomes—i.e., on the basis of considerations that focus on the greatest amount of good and least amount of harm for the greatest number of people. However, this stand in direct conflict with Article 1 and Article 25(1) of the *Universal Declaration of Human Rights* which, respectively, stipulate that

> Article 1: “All human beings are born free and equal in dignity and rights....”

and

> Article 25(1) “[E]veryone has the right to a standard of living adequate for the health and well-being of himself and of his family, including food, clothing, housing and medical care and necessary social services...”

In and by itself, of course, these Articles do not militate against using a utilitarian approach when analysing whether there is a right to health care in general or on a particular occasion. After all, the *Declaration* is the result of a political process and as such holds only within that context and for those who subscribe to it. In and by itself, therefore, this critique lacks universal applicability and validity.

However, the *Declaration* has an underlying logic that validates these clauses objectively speaking. It is a matter of logic that entities that are the same should be treated in the same way.[53] Therefore all human beings or persons, insofar as they are human beings or persons, have the same rights and should be treated the same

---

[51] Sassi (2006).

[52] La Puma and Lawlor (1990), Prieto and Sacristán (2003).

[53] Andrews (2002), Hilbert and Ackermann (1928).

way, and any differences in the social utility of such treatment are irrelevant with respect to the existence of the right in question. This immediately entails Article 25(1): namely, that if any person has a right to medical services or health care, then all people have a right to health care simply because, considered as persons, they are all the same. Consequently it would be inappropriate to decide whether there is a right to health care on the basis of which people are affected, of how many people are affected, or of the practical outcome of recognizing such a right.

Of course in opposition to this one might argue that the sub-clause of this inferences—"insofar as they are human beings"—invalidates this inference because it ignores the fact that human beings differ in their status, their health care needs and in other respects. Consequently it could be replied that it is legitimate to conclude that not all persons have the same right to health care. So, for example, people may differ in their social importance, in their ability to contribute to what is identified as the good, in the cost that would be involved in providing for their health care needs, etc.

However, such a reply would not be a trenchant rejoinder because it would commit a logical error. More specifically, it would conflate the existence of a right with the comparative strength of that right. While the differences in the social positions of individuals, in their ability to contribute to what is identified as the good, or in the cost that would be involved in providing for their health care needs, etc. would indeed be differences between the individuals beings, and while these might make a difference in the relative strength of their rights to health care, this would not affect the fact that they do not differ as human persons and therefore have the same right. It would merely affect the strength of their right-claims—which is logically a different matter.[54]

1.3.6.4 Deontological Ethics

This brings the discussion to deontological ethics as a possible basis for analyzing the right to health care.

Deontological ethical theories are normative ethical theories which contend that the morality of a policy or action is a function of whether that policy or action is consistent with a series of person-centred rules or principles. In other words unlike utilitarian theories, deontological ethical theories do not evaluate policies and action the basis of outcome—e.g., the greatest good for the greatest number—but in terms of considerations that are grounded in what it is to be a person, where the notion of a person is understood as that of an entity that has the capacity for sapient cognitive awareness and volition. Consequently, if such an approach is used as the basis of an analysis of the right to health care, it will frame that analysis without pointing to pragmatic, social, traditional or religious parameters. Not surprisingly, also, a

[54] Moreover, it would make a difference in the relative strength of their rights only if—and to the degree that—the difference in such factors affects the possibility of fulfilling the competing rights in question. For more on this, see *infra*. Chaps. 4 and 5.

deontological approach to ethics claims objective validity because, while it does not ignore the impact that demographic, geographic and temporal factors, etc. may have in specific cases, these considerations do not affect the overall formwork of what is ethically appropriate but only how the framework expresses itself in a given case.

At the same time, as is the case with all other approaches to ethics, deontological theories come in different versions. Three of these in particular have attracted much attention. One of them dispenses entirely with all rules except one, which is more or less procedural in nature: It insists that the way to determine whether an action or policy is ethically acceptable is to generalize the theme of the action or policy into a rule that captures its logic—in other words, to generalize its maxim—and see whether it can function as a universal law without contradiction or doing away with the injunction itself. This is called the *universalizability criterion*. It was most famously expressed by Immanuel Kant in what he called the *categorical imperative*: "Act only according to that maxim by which you can at the same time will that it should become a universal law."[55]

By contrast, what is sometime called a pluralistic deontological theory identifies several rules that are expressed as principles—for instance, as the Principle of Autonomy, the Principle of Equality, and the Principle of Beneficence—and analyses how well the policies or actions in question operationalize them or are in keeping with them. The third approach argues that the principles that are enunciated by the pluralistic deontological approach do not hold absolutely and without exception but only *prima facie*—which is to say, that they are only abstract and general approximations that serve as guides. Their actual implications depend on the facts of the case and hence are context dependent, where the context includes such things as personal, societal and material factors. Only when all of these are taken into account can one say whether a right or duty actually holds or has been followed. Clearly, such a deontological ethical approach agrees with a feminist approach to ethics in this regard.[56]

Like the other approaches to ethics that have been discussed so far, deontological approaches are not without problems when dealing with the right to health care. One can here distinguish between problems of use and problems of specification. Problems of use do not touch the coherence and validity of the theory itself but the usability of its theoretical constraints in actual practice. For instance—and this applies particularly to the Kantian approach—while the theory in question is coherent and may be usable at the policy level, it is unlikely that it can provide useful ethical guidance at the hands-on level. For instance, it is not clear how such an approach can realistically be used in the emergency-room setting where competing right claims have to be resolved within a short time frame and the decision of whom to treat has to be made by professionals whose training lies in health care and not philosophy or logic.

[55] Kant (1993 [1785]). Kant gave several other formulations of his categorical imperative—three more, to be exact—and he claimed that they were all equivalent. Cf. Paton (1947).

[56] See *supra*, p. 9.

A much more serious problem—because it goes to the very validity and structure of their conceptual formworks—is the fact that unless pluralistic deontological approaches can show that the rules or principles that they promulgate are grounded in what it is to be a person, their claim to objective and universal validity is nothing more than a claim—in which case their ethical pronouncements reduce to mere *fiats* or authoritative assertions.

Further issues that are faced by pluralistic deontological theories include the problem of how to resolve conflicts that arise when the rules or principles that structure their frameworks as theories are applied in actual practice, and even how to reconcile these rules or principles themselves with each other. An example of the first of these is illustrated by the fact that the Principle of Autonomy—which is to the effect that all persons have the right to self-determination—raises the question of how far this right extends when different persons make autonomous claims that conflict with each: for instance when, during the COV-19 epidemic, a person may insist on the right not to wear a mask whereas other persons in the same society insist that they do not want to be exposed to the possibility of becoming infected by the breathing of that unprotected persons, or a treating physician may insist on the right not to run the risk of infection from the droplets that spew from the individual's mouth or nose when they cough or even when they breathe.

An example of the second problem is illustrated by the question of how to reconcile the Principle of Beneficence (which is to the effect that people have a duty to advance the good of other people) when advancing that good produces harm to third parties: i.e., when it conflicts with the Principle of Nonmaleficence (which is to the effect that everyone has a duty not to harm other persons). Thus, according to the Principle of Beneficence physicians have a duty to act in the best interest of their patients, and in certain circumstances that means putting a patient on life support in order to save their life. However, by the patient's estimation that would involve doing harm—and thus would be a violation of the Principle of Nonmaleficence—because the patient considers being kept alive as torture, and wishes nothing more than to have assistance in dying.

Another critique that has been levelled against deontological approaches is that they seem to disregard entirely the consequences of the policies or actions that they apparently validate. Therefore such approaches may be fine in theory but unusable in practice, because in real life consequences always matter.

However, when closely considered these critiques neither impugn the logical consistency of deontological ethical approaches in general nor show that they are inappropriate when analysing the right to health care. What they show is that if a deontological approach is used, then it has to spell out how its principles or rules are derived, that they are consistent with each other, and that they can be applied without conflict in actual practice. Moreover, none of the critiques reject what would be central to a deontologically grounded analysis of the question whether there is a right to health and what shape this right might take: namely, the focus on persons as persons. And that, after all, is what health care is all about: the health-oriented treatment of persons. It is for these reasons that a deontological approach will be used in the rest of this study. This of course means that a separate section has to be devoted to spelling out the details of this deontological approach because, as the preceding comments quite correctly indicate, what has been presented so far amounts to no more than a sketch.

1.4 The Nature of Health

However, before providing such an account, a further matter requires attention: namely, the notion of health itself. Unless that term "health" is clearly defined and understood, the very notion of health care itself will be ambiguous, and what should or should not be included under its ambit will remain unclear.

The task of defining 'health', however, is much like the task of defining "number". Thus, there are various kinds of numbers: natural numbers, odd numbers, even numbers, finite numbers, infinite numbers, rational numbers, irrational numbers, real numbers, negative numbers, positive numbers, imaginary numbers, and so on. Therefore, in order to capture what is characteristic of numbers as numbers and to be applicable to all numbers, the definition of "number" has to capture the logic of what it is to be a number irrespective of their individual kinds.

The same things holds true for the concept of health. There are various kinds of health, and some of these are as distinct as kinds of numbers. Within the field of health care alone, as the term is commonly understood, there is care that focuses on physical health as opposed to mental health, psycho-social health, pediatric health, geriatric health, genetic health, etc.

Not only that: the term "health" itself has been defined in various ways. For instance, in 1948 it was defined by the World Health Organization (WHO) as "a state of complete physical, mental and social well-being and not merely the absence of disease or infirmity,"[57] where this definition was revised in 1984 as "The extent to which an individual or group is able, on the one hand, to develop and satisfy needs; and, on the other hand, to change or cope with the environment. Health is therefore seen as a resource for everyday life, not the objective of living; it is seen as a positive concept emphasizing social and personal resources, as well as physical capacities."[58] It has also been defined in statistical and species-relative terms as the "statistically normal biological function with … typical efficiency" for a given species.[59] Further definitions include "a person's psychophysiological ability to act or respond (in a way that is supportive of the person's goals, projects or aspirations) in a wide variety of situations;"[60] "the ability to achieve … vital goals;"[61] as well as "a state characterized by anatomic, physiologic, and psychological integrity; ability to perform personally valued family, work, and community roles; ability to deal with physical, biological,

[57] World Health Organization, Constitution (1948).

[58] World Health Organisation (1992).

[59] Boors (1977).

[60] Whitbeck (1981).

[61] Nordenfelt (1995).

psychological, and social stress."[62] It has even been defined in normative terms that depend on the philosophical view that one has of human nature.[63]

This variety of definitions notwithstanding—and here are still others—several things have to be kept in mind when defining the notion of health if it is to be used in an ethically grounded analysis of the right to health care. The first is that the fact that statistical normalcy is not the same as being healthy. For instance, it may be statistically normal for a given population to suffer from thalassemia or to carry the thalassemia gene, but that does not mean that thalassemia itself is not a health condition or that the individuals in question are healthy.[64]

Second, a purely social or cultural perspective[65] cannot be used in an analysis of the right to health care because it lacks reference to the fact that human beings are animals, and that the condition and functioning of their bodies falls squarely under the rubric of health. For the same reason, a definition whose focus centres in a particular philosophical conception of what it is to be a person[66] would be incomplete because, again, it would not take into account the fact that human beings are animals, which makes the biological functioning of their bodies an integral aspect of their health status.

At the same time, a definition of "health" that does not acknowledge that human beings are more than biological entities—that they have the capacity for cognitive awareness and volition, and that they live in societies—would also be incomplete. Therefore any acceptable definition of "health" must span the biological, cognitive, social and psychological spectrum.

Third, an acceptable definition of "health" has to acknowledge that the environment has an effect on whether a biological organism—and that includes human beings—is healthy. More specifically, the fact that an organism is healthy in one environment does not entail that it will be healthy if that environment changes or the individual migrates to a different environment. So, for instance, the fact that someone's lung capacity and ability to extract oxygen from the air functions well at sea level or in a smog-free environment does not mean that if the oxygen concentration shifts to what is found at high altitudes such as the Andes or the Himalayas, or if air pollutants approach or even exceed the concentration that is found in heavily polluted cities like Hotan or Raiwind, the individual will be healthy.

Fourth, the physiological status and functioning of human beings changes over time as they progress through different stages of their life cycle, and what is healthy at one stage may not be what is healthy at another. For instance the pharmacokinetic functioning of a two-year old is different from that of a fifty-year old,[67] and what is healthy at one stage is not necessarily what is healthy at another.

---

62 Stokes et al. (1982).

63 Engelhardt (1975).

64 Weatherall (2018).

65 Kramer (2002).

66 Nordenfelt, *op cit*.

67 Ginsberg et al. (2002).

At first glance, then, the following functional definition of 'health' seems to be appropriate because it seems to acknowledge these considerations:

> Health is a state of homeostatic balance that allows the organism to survive, mature and function in the environment in which they are embedded.

It correctly notes that health is a functional notion, and that it is context dependent, is not invariant for all the life-stages of a biological entity that undergoes maturational development, and it can accommodate considerations that involve group embedding.

However, it is insufficient for the present study because it captures only the purely physiological aspects of human beings. As has already been emphasized, human beings also have sapient, cognitive and volitional capacities. Therefore, it seems appropriate to amend the definition as follows:

> Health is a state of homeostatic balance that allows the organism to survive, mature and function in a specific-specific manner in the environment in which they are embedded

It should be clear that this amended definition allows for an extension of the notion of health to the human sphere where human beings are understood as persons in the sense defined in the preceding parts of this preliminary discussion. It should also be clear that it allows for a corresponding definition of illness or disease, while at the same time being neutral as to why or how an absence of health—i.e., illness or disease—is brought about. Moreover, it spans the range of services that are provided in the name of health care, inclusive of physiological and psycho-social services. At the same time, it does not insist that the relevant capacities must be exercised, nor does it insist that the individuals—specifically, human beings—must see themselves as sexual beings of a particular sort or that they have to structure the nature of their social interactions along particular lines in order to be considered healthy. In other words, it is neutral with respect to gender identity and does not characterise it as a matter of health. Finally, it is neutral on the question whether there is a right to health care and, equally as importantly, whether that right has to be exercised by a putative right-holder.

1.5 Conclusion

The aim of the preceding discussion has been to clarify how terms that are central to an analysis of the right to health care will be understood in the chapters that follow, and why they will be understood in this particular way. At the same time, however, while the preceding may have clarified how the terms will be understood and has identified the type of ethical approach that will form the framework for the analysis—namely, a pluralistic deontological approach—and why this approach will be used, it has not spelled out what precisely this means in conceptual terms. What is lacking so far is a precise indication of the actual principles that structure the analysis, nor has it indicated how these principles are derived from the concept of a person—something that is integral to a discussion of rights. Nor, finally, has it clarified the logical interrelationship between these principles—indeed, their consistency with each other—and how they fit in with the points that were raised in the sketch of deontological approaches above. Before proceeding with the analysis of the right to health care, these lacunae will have to be filled since otherwise the discussion would be beset by confusion, ambiguity and unclarity, the preceding definitional discussion notwithstanding. The next chapter, therefore, will be devoted to filling this gap. The analysis proper will then developed in the subsequent chapters.

References

Althaus, F.A. 1997. Female circumcision: rite of passage or violation of rights? *International Perspectives on Sexual and Reproductive Health* 23: 3, 130.

Andrews, Peter B. 2002. *An Introduction to Mathematical Logic and Type Theory: To Truth Through Proof*, 2nd ed. Berlin: Kluwer Academic Publishers.

Audi, R. 2004. *The Good in the Right: A Theory of Intuition and Intrinsic Value*. Principia Ethica: Princeton University Press; Moore GE.

Badhwar, N.K. 2014. *Well-Being: Happiness in a Worthwhile Life*. Oxford University Press.

Baehr, A., ed. 2004. *Varieties of Feminist Liberalism*. Lanham: Rowman and Littlefield Publishers.

Bentham, J. 2009. *An Introduction to the Principles of Morals and Legislation*. Dover Philosophical Classics. Dover Publications.

Boors, C. 1977. Health as a theoretical concept. *Philosophy of Science* 44 (4): 542–573.

Brandt, R.B. 1979. *A Theory of the Good and the Right*. Oxford Clarendon Press.

Chappell, T. 2014. *Knowing What to Do*. Oxford University Press.

Dawson, D. 2009. Conflicting stories of virtue in UK healthcare: Bringing together organisational studies and ethics. *Business Ethics: A European Review*18 (2): 95–109.

Dillon, R.S. 2020. *Feminist Approaches to Virtue Ethics*. Oxford University Press.

Drane, J.F. 1995. *Becoming a Good Doctor: The Place of Virtues and Character in Medical Ethics*. Kansas: Sheed and Ward.

Ronald M. Dworkin. 1986. *Law's Empire*. Cambridge: Harvard University Press.

Eggleston, B. 2014. Act utilitarianism. In *The Cambridge Companion to Utilitarianism*, ed. by Ben Eggleston and Dale E. Miller, 125–145. Cambridge University Press.

Emon, A.M, Levering, M., Novak D. 2015. *Natural Law: A Jewish, Christian, and Islamic Trialogue*. Oxford University Press.

Engelhardt, Jr. HT. 1975. The concepts of health and disease. In *Evaluation and Explanation in the Biological Sciences* ed. Engelhardt and Spicker, 125–141. Dordrecht: Reidel.

Finnis, John. 1980. *Natural Law and Natural Rights*. Oxford: Clarendon Press.

Firestone, S. 1970. *The Dialectic of Sex: The Case For Feminist Revolution*. New York: Morrow.

French, P.A. 1979. The corporation as a moral person. *American Philosophical Quarterly* 16 (3): 297–317.

Gilligan, C. 1982. *In a Different Voice: Psychological Theory and Women's Development*. Harvard University Press.

Ginsberg, G., D. Hattis, B. Sonawane, A. Russ, P. Banati, M. Kozlak, S. Smolenski, and R. Goble. 2002. Evaluation of child/adult pharmacokinetic differences from a database derived from the therapeutic drug literature. *Toxicological Sciences* 66 (2): 185–200.

Gough, J.W. 1936. *The Social Contract*. Oxford: Clarendon Press.

Haakonssen, K. 1996. *Natural Law and Moral Philosophy: From Grotius to the Scottish Enlightenment*. Cambridge, UK: Cambridge University Press.

Hare, R.M. 1981. *Moral Thinking: Its Levels, Method, and Point*. Oxford, England: Clarendon Press; Singer, *op. cit*.

Hay, C. 2013. *Kantianism, Liberalism, and Feminism: Resisting Oppression.* London: Palgrave-Macmillan.

Held, V. 2006. *The Ethics of Care: Personal, Political, and Global*, 2nd ed. Oxford New York: Oxford University Press.

Hess, K.M. 2014. The free will of corporations (and other collectives). *Philosophical Studies* 168 (1): 241–242.

Hilbert, D., and W. Ackermann. 1928. *Grundzüge der theoretischen Logik (Principles of Mathematical Logic).* Springer.

Hodgkinson, C. 1996. *Administrative Philosophy: Values and Motivations in Administrative Life.* Pergamon.

Jaggar, A.M., and S.R. Bordo (eds). 1992. *Gender/body/knowledge: Feminist Reconstructions of Being and Knowing.* Rutgers University Press.

Kant, Immanuel. 1993 [1785]. *Grounding for the Metaphysics of Morals*, trans. and ed. Ellington, James W., 3rd ed. Hackett. p. 30. Kant gave several other formulations of his categorical imperative—three more, to be exact—and he claimed that they were all equivalent.

Kramer, R.M. 2002. When paranoia makes sense. *Harvard Business Review* 8:7, 62-69.

La Puma, J., and E.F. Lawlor. 1990. Quality-adjusted life-years. Ethical implications for physicians and policymakers. *Journal of the American Medical Association* 263 (21): 2917–2921.

Lindemann, H. 2005. *An Invitation to Feminist Ethics.* McGraw-Hill, *passim.*

List, C., and P. Pettit. 2011. *Group Agency: The Possibility, Design, and Status of Corporate Agents.* New York: Oxford University Press.

Lon, L. 1964. *Fuller, The Morality of Law.* Revised Edition. New Haven: Yale University Press.

Mandal, J., D.K. Ponnambath, and S.C. Parija. 2016. Utilitarian and deontological ethics in medicine. *Trop Parasitol* 6 (1): 5–7.

Moore, G.E. 1912. *Ethics.* Williams and Norgate.

Mill, J.S. 1998. Crisp, Roger (ed.). Utilitarianism. Oxford University Press.

Nordenfelt, L.Y. 1995. *On the nature of Health: An Action-Theoretic Approach.* Intoduction: Springer.

Nussbaum, M. 1999. *Sex and Social Justice.* Oxford: Oxford U Press.

Parfit, D. 1986. *Reasons and Persons.* Oxford Paperbacks.

Paton, A.J. 1947. *The Categorical Imperative: A Study in Kant's Moral Philosophy.* London: Hutchinson.

Pellegrino, E.T.D. 1993. *The Virtues in Clinical Practice.* New York: Oxford University Press.

Toon, P. 2014. *A Flourishing Practice?* London: Royal College of General Practitioners.

Prieto, L., and J.A. Sacristán. 2003. Problems and solutions in calculating quality-adjusted life years (QALYs). BMC: *Health and Quality of Life Outcomes* 1: 80.

Rawls, J. 1971. *A Theory of Justice.* Cambridge, MA: Harvard University Press.

Riley, P. 1982. *Will and Political Legitimacy: A Critical Exposition of Social Contract Theory in Hobbes.* Locke, Rousseau, Kant, and Hegel, Harvard University Press.

Rommen, H.A. 1947 [reprinted 1959]). *The Natural Law: A Study in Legal and Social Philosophy*, trans. Thomas R. Hanley, Herder Book Co.

Rousseau, J.-J. 1762. *The Social Contract, or Principles of Political Right.*

Sassi, F. 2006. Calculating QALYs, comparing QALY and DALY calculations. *Health Policy and Planning.* 21 (5): 402–408.

Sherwin, S. 1991. Abortion through a feminist lense. *Dialogue* 30:3, 327–342.

Sidgwick, H. 1981. *Methods of Ethics*, 7th ed. Hackett Publishing Co.

Singer, P. 2011. *Practical Ethics*, 3rd ed. Cambridge University Press.

Sinhababu, N. 2018. Scalar consequentialism the right way. *Philosophical Studies* 175: 3131–3144.

Slote. 2001. *Morals from Motives.* Oxford: Oxford University Press.

Stokes, J., J. Noren, and S. Shindell. 1982. Definition of terms and concepts applicable to clinical preventive medicine. *Journal of Community Health* 8 (1): 33–41.

Stone-Mediatore, S. 2004. Women's rights and cultural differences. *Studies in Practical Philosophy* 4 (2): 111–133.

Sullivan, R. 2003. Statutory Interpretation in a New Nutshell. *Canadian Bar Review* 82 (1): 51–82.
Teichman, J. 1985. The definition of person. *Philosophy* 6 (232): 75–185.
Tong, M. 2006. *New Perspectives in Healthcare Ethics: An Interdisciplinary and Crosscultural Approach*. Westview Press.
Weatherall, D.J. 2018. The evolving spectrum of the epidemiology of thalassemia. *Hematology/oncology Clinics of North America* 32 (2): 165–175.
Whitbeck, C. 1981. A theory of health. In *Concepts of Health and Disease*, ed. Caplan, A.L., Engenhardt, H.T. Jr., McCatrney J.J, 611–626 at. 620. Adison-Wesley.
Williams, B. 1985. *Ethics and the Limits of Philosophy*, 1. Harvard University Pressp.
Carla, Wilson. 2001. Decolonizing Methodologies: Research and Indigenous Peoples, Social Policy Journal of New Zealand 17 (December 2001), 214–217.
World Health Organisation. 1992. *Basic Documents*, 39th ed. Geneva: WHO. https://apps.who.int/gb/bd/. Accessed 18 May 2020.
World Health Organization, Constitution. 1948. https://www.who.int/about/who-we-are/constitution. Accessed 18 May 2020.
World Health Organization. 2017. *Female genital mutilation*. http://www.who.int/mediacentre/factsheets/fs241/en/.
Zagzebski. 2010. Exemplarist Virtue Theory. *Metaphilosophy* 41(1/2): 41–57.

Chapter 2
Ethical Theory: Persons, Ethical Rules and Principles

Abstract The present chapter expands the notion of personhood to include constructive persons such as societies and institutions in order to validate the relevance of ethical considerations as applied to them. It outlines the ethical principles that will be involved in the subsequent analysis, shows how they logically derived from the notion of person and the relationship of the principles to each other. It thus sets set the stage for the analysis of the right to health care that is presented in the remaining chapters.

Keywords Constructive person · Principles of ethics · Best practices

2.1 Introduction

The aim of the preceding chapter was to clarify how the terms that will be used in the rest of this study will be understood, why they will be understood that way, and why a particular ethical approach—deontological ethics—will form the framework of the analysis. The reason that was given for choosing this approach—and more particularly still, for choosing a pluralistic deontological approach[1]—was that this approach centres in the notion of a person and that, unlike other ethical approaches, its principles can either be derived from or can be logically related to what it is to be a person while at the same time being usable in actual practice. The notion of a person, in turn, was defined as that of entity that has the capacity for cognitive awareness and volition.[2] The task of the present chapter is to outline how and why this is the case, to identify the principles that are involved, to show their logical derivation from the notion of person and their relationship with each other, and thereby to set the stage for the actual analysis that will be presented in the remaining chapters.

---

[1] In the rest of this study, the term "pluralistic" will be taken as understood.
[2] Cf. Chap. 1, pp. 7 f. *supra*.

E.-H. W. Kluge, *The Right to Health Care: Ethical Considerations*, The International Library of Bioethics 92, https://doi.org/10.1007/978-3-030-93838-3_2

However, before actually proceeding with this, it may be appropriate to briefly reemphasize a point that was made in the preceding chapter when clarifying the notion of a person. It was then pointed out that there are different kinds of persons, and that the notion extends beyond the standard case of a living member of the species *Homo sapiens*: in other words, that it encompasses more than individuals who fall into the rubric of *natural person*. More specifically, it was said that groups of individuals who function as a unit and take cognizance of certain facts, have certain values and aims and make decisions may also be considered persons. By way of underwriting this claim, it was pointed out that this is formally recognized ethically, socially as well as legally in the concept of a *corporate person.*[3]

The reason this is important for the present study is that corporate persons—which include not only corporations but also societies and institutions—are just as much subject to ethical evaluation as are natural persons, which in turn means that with due alteration of detail the principles that structure the framework of deontological ethics and that will be detailed in what follows also apply to them. And the reason why this is important is that it is from this that it follows that it is appropriate to subject the health care related policies that are promulgated by a society, by a corporation or by an institution, as well as their actions, to ethical scrutiny and not merely to legal considerations.

2.2 The Principles of Ethics

2.2.1 Rights and Duties

Turning, then, to an outline of the deontological approach to ethics, it may be appropriate to preface it with a brief remark about rights and duties.

As was indicated in the previous chapter,[4] rights and duties are correlatives. This means that when there is a right there is a corresponding duty, and that when there is a duty there is a corresponding right. It was also suggested that rights in the ethical sense properly belong only to persons; and further, that if someone has a right then they can choose whether to exercise that right. They may exercise that right but need not do so. It was also suggested that they may forego the exercise of that right in general, or they may forego to exercise it only on a given occasion. This becomes particularly important in the case of health care because, as will be argued in more detail later, individuals may forego the exercise of their right to health care on a particular occasion—for instance, when there is a scarcity of resources or

[3] *Concise Encyclopedia of Business Ethics,* "Corporate Personhood;" https://conciseencyclopedia.org/entries/corporate-personhood/#:~:text=Corporate%20personhood%20is%20the%20ethical,human%20beings%20associated%20with%20them.&text=Like%20human%20persons%2C%20corporations%20have,in%20pursuit%20of%20those%20goals.

[4] Chapter 1, p. 1, *supra.*

when the exercise of their right would conflict with the exercise of another person's right—without thereby abandoning their right to health care in general.

It was also suggested that value considerations apply both in the case of rights and of duties: namely, with respect to their exercise. This is not to say that whether someone has a right or a duty is a function of their choice or of their values. That is functionally determined by the state of affairs in question and the ethical principles that apply in that situation. Values, however, enter with respect to the question whether they wish to exercise their right or act in accordance with their duty.

However, rights and duties differ in that a right-holder may legitimately forego the exercise of their right if they wish to do so—which is usually a function of their values—but that a duty-holder's values have no relevance with respect to whether they may legitimately forego acting in accordance with their duty. All other things being equal, they should act in accordance with that duty. Of course whether they in fact do so is not a foregone conclusion. In other words, having a duty is not like a causal relationship where, given the antecedent, the consequent follows. Rather, the duty-holder may choose to act in accordance with that duty or may choose not to do so—which is a function of the values that they hold. Therefore choice enters here as well. However, whereas in the case of rights the fact of choosing not to exercise that right does not affect the ethical stance of the right-holder, the fact of choosing not to act in accordance with their duty does affect the ethical stance of the duty-holder: Not to act in accordance with one's duties is to be guilty of an ethical failure.

The upshot of this, therefore, is that neither in the case of rights nor in the case of duties is the relationship between them and what in fact happens like a causal relationship where, given the antecedent condition, the result automatically follows. It is always a matter of choice and values. This is why acting ethically is always a matter of choice.

2.2.2 *The Principle of Autonomy*

The domain of ethics, then, is the domain of choice: One cannot be a moral agent if one has no capacity for volition or if one cannot choose.[5] Therefore it should come as no surprise that one of the most fundamental principles of ethics centres in the notion of choice itself. This is the *Principle of Autonomy*.

Every person has the right to self-determination.

However, that is not the end of the matter because the self-determination of one person may conflict with the self-determination of another. Without a way of resolving such a conflict, the Principle of Autonomy would set up irresolvable conflicts in the framework of ethics itself and thereby make that framework logically incoherent.

---

[5] Tannsjo (1999).

The clue on how this is resolved lies in the words of the noted American legal scholar Zechariah Chafee, who put it in colloquial terms as follows[6]:

> Your right to swing your arms ends just where the other man's nose begins.

The point of this statement was that autonomy is not the same as licence. Autonomy is the right to self-determination within the realm of what is consistent with the right to autonomy of other persons, whereas licence is the exercise of self-determination without such limitation.

That is to say, a framework that purports to govern what are considered to be rights—and, by extension, duties—and that determines how actions may legitimately be conducted has to have a logically consistent structure since otherwise the very notions of right and constraint lose all ethically binding force. This means that if the framework of ethics determines what are rights and what are duties, and if this is a function of the principles of ethics, then the principles themselves must contain conditions that allow their implementation without mutual contradiction. This, in turn, means that the principles have to have inherent and logically grounded limitations, and that this should be reflected in an acceptable formulation of the principles themselves.

However, as was just pointed out, the untrammeled exercise of autonomy—in other words, construing autonomy as licence—would violate this condition. Therefore the Principle of Autonomy carries certain logically grounded limits within itself, which of course means that this should be reflected in a proper formulation of the Principle. The initial statement of the Principle, therefore, has to be amended as follows:

> Every person has the right to self-determination subject to the equal and competing rights of other persons.

It is perhaps worth emphasising that this does not mean that this limitation of autonomy is a matter of convenience or practicality. It is strictly a matter of logic. Therefore if such a limitation cannot be established on purely logical grounds, it is not a legitimate limitation.

It is also worth reemphasizing that the legitimacy of such a limitation does not fall solely within the realm of autonomy alone: i.e., that it is only when an exercise of autonomy conflicts with another exercise of autonomy that limitation is appropriate. The requirement of logical consistency of the framework of ethics itself entails a corresponding limitation within the realm of ethics as a whole. This means that the rights and duties that are legitimated by any ethical principles are legitimated only *prima facie*, and that their full exercise is subject to the limitation that they not set up irresolvable conflicts within the domain of ethics.[7]

---

[6] Chafee (1919).

[7] It should be noted that this condition is different from the condition of *prima facie* validity mentioned in Chapter I. That condition focused on the factual parameters of the specific implementation of a right or duty. In other words, it focused on pragmatic contexts and did not necessarily involve a logical conflict of duties and rights. The condition that was just outlined focuses on logical conflicts.

2.2.3 *Principle of Impossibility*

Interestingly enough, the fact that the framework of ethics has to be logically consistent can itself be expressed in the form of another principle which, while strictly speaking not a principle of ethics, applies to all such principles and, indeed, to all rights and duties that derive from them. It is the Principle of Impossibility.

Thus, it does not follow that when someone exercises a right that the party who has the corresponding duty but does not fulfil that right is automatically guilty of an ethical fault. So, for instance, if someone has a right to health care but whoever has the duty to provide that care does not do so, it does not automatically follow that the duty-holder is therefore ethically guilty. The resources that might be necessary to provide that care may not exist or may be otherwise unavailable or the individual who otherwise would have the corresponding obligation may themselves be incapacitated, in which case and under such circumstances the right itself lapses.

The Principle of Impossibility, therefore, functions as an overarching principle that conditions not merely the Principles of Ethics themselves, but all rights and duties. Going back to Chap. 1, the reason for this lies in the logical nature of rights and duties themselves: As was then said, while it possible to articulate in a grammatically unexceptional manner rights and duties that do not satisfy the Principle of Impossibility, the fact that these would involve a contradiction entails that, logically speaking, doing they would be conceptually meaningless.[8]

The point of what was just outlined, then, can be captured by what is generally called the *Principle of Impossibility*:

> All rights and duties hold subject to the condition that it is possible to fulfil them under the circumstances that obtain.

The validity of this principle is recognized in not merely in ethics but also in law. It was already an integral part of Roman law,[9] is fundamental to modern jurisprudence and has found reflection in most legal jurisdictions. To cite but a few specific examples of this, it finds reflection in the laws of India,[10] Japan,[11] Russia[12] and Thailand[13] as well as in the laws of Commonwealth countries that follow the legal traditions of the United Kingdom, where it was fundamental to the leading case of *Taylor v. Caldwell*.[14] It is also accepted in international law in the context of treaty

---

[8] As will become important later when dealing with allocation of resources, impossibility is not the same as futility. Futility looks at whether the relevant action will be successful in producing the intended or expected result. Thus, it would be futile to perform a colectomy on someone dying of inoperable metastasized cancer of the colon which has spread through the whole body and neither chemotherapy nor radiation have proved or will prove effective. It would be futile because it would not change the fact of impending death.

[9] Justinian, *Digest* 50. 18. 185.

[10] India, *Contract Act* 1972s. 56.

[11] Japan, *Civil Code* art. 415 is generally interpreted as having this as an underlying principle.

[12] Sadikov (2015).

[13] See Chen-Wishart et al. (2016).

[14] *Taylor v Caldwell* [1863] EWHC QB J1.

obligations.[15] Returning, however, to ethics, the Principle logically structures the whole framework of deontological ethics, whether that be understood in a Kantian or a pluralistic fashion.[16] Therefore it applies to all rights and duties as these will be discussed in the rest of this study.

At the same time, it is important to note that the concept of impossibility as it is understood here should not be construed in the sense of inconvenience or inopportuneness relative to the implementation of a right or the execution of a duty. If there is a right to health care and a corresponding duty, then the fact that it may be expensive, difficult or inopportune to provide such care or that it involves problems for whoever has that duty does not mean that the right lapses or that the duty is thereby rendered null and void.

Moreover, it should also be understood that if the impossibility in question is the result of an inappropriate action or the failure to act appropriately on part of the individual who otherwise would have the relevant duty, the Principle does not apply either. The reason for this lies in the fact that in such cases the impossibility is not a function of the content of obligation itself but of a failure on part of the duty-holder to adhere to their duty. In other words, it would not be a matter of logic but of decision and implementation. Consequently, follows that if someone has right to health care and the party who has the corresponding obligation arranges matters in such a way that it becomes financially prohibitive or otherwise makes it impossible to provide that care—perhaps because of other arrangements that the duty-holder has made—then the duty to provide the care still remains, and the duty-holder is ethically culpable if they do not provide that care.

The initial formulation of the Principle of Impossibility should therefore be amended as follows:

> All rights and duties hold subject to the condition that it is possible to meet them under the circumstances that obtain, where the fact of impossibility lies in the nature of the rights or duties or of the situation but is not a result of the voluntary actions of the relevant party in question.

2.2.3.1 The Doctrine of Necessity

But again, that is not the end of the matter. The fact is that fulfilling an otherwise legitimate right may make it impossible to fulfil another competing right, where the second right is logically more basic than or logically prior to the first. So, for instance, meeting an otherwise legitimate right to health care in the best way possible may involve using resources that are necessary to defray the basic health care needs of other individuals who also have a right to health care. Likewise, if a health care professional did not break the obligation of patient confidentiality—which latter is based on the right to privacy[17]—and inform the relevant parties of the fact that the

[15] See Reuter (1985).

[16] The qualifier "pluralistic" will be taken as understood in the rest of what follows.

[17] Chapter 6 pp. 129 ff.

patient suffers from a serious communicable disease, the health care professional would put other individuals and society at risk. Under such circumstances, the Principle of Impossibility still applies, but in a modified form, which is to say under the guise of the *Doctrine of Necessity*[18]:

> If a right claim cannot be met without at the same time making it impossible to fulfil a more basic and competing right, then the right in question does not lose its force but need only be fulfilled to the degree that is compatible with the fulfilment of that more basic right.

This, of course, raises the question whether the notion of the basicness or priority of rights—in other words, whether the question of the ranking of rights relative to each other—can be established in logical terms without falling prey to purely pragmatic or legal considerations. The issue is of some importance because such ranking plays a crucial role at the hands-on level of health care when it comes to ethically justifying triage algorithms, as well as at the policy level when it comes to justifying the ordering of spending priorities. For the time being, however, the matter will be left here. It will be examined further in the next chapter.

2.2.4 *Principle of Equality*

Another principle that is grounded in logic is: the *Principle of Equality*:

> All persons, insofar as they are persons, are equal and should be treated in the same way.

The reason that it is grounded in logic is that if there is no relevant qualitative difference between distinct entities, then differential treatment becomes a matter of arbitrary decision. This holds true no matter what the nature of the entities in question, and it is captured in the equation

$$\text{If } x = y, \text{ then } f(x) = f(y)$$

where x and y *stand* for any entities whatever.

However—and this addresses the issue of what is mean by "relevant qualitative difference"—while persons insofar as they are persons are the same and therefore should be treated in the same way simply as a matter of logic, this holds true only insofar as they are persons. In real life, however, persons are embedded in a social context. Moreover, they may differ physically as well as psychologically in their attributes, which means that it may be legitimate to treat them differently on such a basis.

Still, in many ways what counts as a relevant difference is a function of the conceptual framework that is used to judge that the relevant entities are in fact different.

[18] It is generally held that the first discussion of this is found in the writings of the medieval jurist Henry de Bracton. (c. 1235) *De legibus et consuetudinibus regni Angliae*, which was founded on Roman law.

For instance, differences in musical aptitude or musical achievement do count as differences if the framework is that of music but do not count as differences if the framework is that of medicine. Likewise, sexual differences count if the framework is that of reproductive capacity but do not count as differences if the framework is that of administrative capacity. It therefore follows that if the framework is that of ethics, then what count as differences must be ethically relevant characteristics, which in turn means that they must be differences in the natures of the individuals or in the situations in which they are embedded. Consequently the initial formulation of the Principle of Equality is importantly incomplete because it ignores this logically relevant fact. Its proper formulation therefore is as follows:

> All persons, insofar as they are persons, are equal and should be treated the in the same way. Exceptions to this must always be based on ethically relevant differences in the natures or the situations of the persons in question.

This of course raises the question of would count as an ethically relevant difference in the nature or situation of a given person, and what effect such a difference has on the rights that they would otherwise have.

By way of answering the question, it may be useful to give an example that is drawn from the interface between the notion of personhood and that of a right. As was just pointed out, all persons have the right of self-determination. However, not all persons can exercise that right because some lack the ability to apprehend and fully understand the choices with which they are faced. Very young children and mentally impaired individuals are cases in point, as are persons who are in a coma or under a general anaesthetic. However, that does not mean that they lose their right to self-determination. It merely means that they must be treated differently when handling that right. In other words, it means that due accommodation must be made, since they still satisfy the criterion for having the right to self-determination

It may be useful to expand on this because it will become important later. To put it first in general terms, there is a difference between the existence of a right and the ability to exercise that right. Whether an individual has a right does not depend on whether the individual can exercise the right but on whether the individual satisfies the criterion for ascribing the right to them.

Applying this distinction to the case at issue, it means that the right to self-determination belongs to individuals as long as they are persons. Whether they are able to exercise that right or how well they are able to do so is another question. As to satisfying the criterion for personhood, as was pointed out before, persons are biological entities who have the constitutional capacity for cognitive awareness and volition. And as the Ad Hoc Committee of the Harvard Medical School to Examine the Definition of Brain Death pointed out, the constitutional capacity for cognitive awareness and volition is grounded in the possession of a functioning brain.[19] Therefore as long as the individuals in question have a functioning brain, they will count as persons, and as long as they count as persons they satisfy the

---

[19] *A definition of irreversible coma.* Report of the Ad Hoc Committee of the Harvard Medical School to Examine the Definition of Brain Death; AMA, 1968 Aug 5;205(6):337–40.

criterion for ascribing the right to self-determination to them. Consequently, their inability to exercise their right to self-determination constitutes an ethically relevant difference—which, given the Principle of Equality as stated above, does not mean that they lose their right but that they must be treated differently while acknowledging that they still have that right. Consequently, the Principle of Equality entails that in their case due accommodation must be made—which, as will be spelled out more fully in the chapters that follow—means that a duly empowered substitute decision maker must be provided who will exercise their right on their behalf.[20]

Another sort of example is provided by situations in which individuals meet the cognitive and volitional conditions that are necessary for them to exercise their rights in their own behalf but where the situation in which they find themselves prevents the implementation of their decisions. For instance, not all people live in locations where there are institutions that can provide health care. Small hamlets in remote areas would here be a case in point. Similarly, not all geographic locations have professional specialists who can provide the services that are necessary to fulfil an individual's otherwise extant right to health care.

That, however, does not mean that such individuals lose their right to health care. As will be explained more fully in the next chapter, the right to health is entailed simply by their being persons who are members of society. Their situation merely means that the location would make it very expensive to place health care institutions in that place or to locate the necessary professionals in that location. However, it would not be impossible to do so—which means that the Principle of Impossibility would not be engaged and hence, because they satisfy the criterion for having the right to health care, they still have the right to health care It merely means that due accommodation must be made to deal with this difficulty because, as was said a moment ago, the right to health care belongs to such individuals simply because they fulfil the criterion of being persons who are members of society.

At the same time, their location does constitute an ethically relevant difference. The Principle of Equality therefore entails that things like appropriate transportation should be made available to such individuals in order for them to be able to reach the locations where the relevant care can be provided, or that other means should be put in place in order to satisfy their right to health care—for example, by instituting measures such as eHealth and telemedicine.[21]

To put it again in general terms, things would be different if overcoming the pragmatic difficulties in satisfying a right would make it impossible to fulfil an otherwise supervenient right. For instance, that would be the case if providing the best possible treatment modality for a given condition would be so expensive that it would take up all the resources that would be required to provide health care to other members of society. At the same time, however, that would still not mean that the right to health care itself would not be effective. It would only mean that the duty

---

[20] For more on this, see Chap. 8, pp. 175 ff.

[21] For more on this, see Chap. 4, "Allocation of Resources."

to provide health care would be satisfied if an appropriate treatment modality other than the best possible treatment morality was provided.[22]

These, then are some examples of ethically relevant differences in the nature, status or situation of persons that would justify differential treatment without violating the Principle of Equality. The Principle and its modifications will play a crucial role in the discussion of triage and discriminatory resource allocation in the chapters that follow, as well as when dealing with the fact that distinct privacy considerations may be ethically justified for different persons if these persons are ethically distinct in their status or their roles.

2.2.5 *Principle of Beneficence*

Another principle of pluralistic deontological ethics is the *Principle of Beneficence.* An initial formulation of this Principle might go as follows:

> Everyone has a duty to advance the good of others insofar it they can do so without undue risk to themselves.

where what counts as good is whatever serves to preserve or advance the existence or wellbeing of the relevant individual.

At first glance, this principle lacks logical justification in what it is to be a person, and therefore does not meet the condition that was stated at the beginning of this chapter as a requirement for the objective validity of ethical principles.

However here, as in many other cases, this first impression is deceiving because the conceptual net of the analysis is not cast widely enough. In this case, the conceptual net should be extended to include both the clause "without undue risk to themselves" which limits the ambit of the principle, as well as the relationship between societies and persons.

As to the first, here the point to keep in mind is that the principle states a general condition that applies to all persons. Societies, of course, consist of persons. Therefore if the duty that is enunciated by the principle were to be followed without safeguarding the existence or wellbeing of the agents that implement the principle, society itself would cease to exist and function because its constituent members would cease to exist or not be well enough to function. Consequently, without this limiting clause, following the principle would be self-defeating because the agents who are supposed to act in accordance with the principle would not be around or be able to implement it. Not to put too fine a point on it, the Principle of Beneficence would violate the Principle of Impossibility.

The second thing that should here be kept in mind—namely, the relationship between societies and persons—is that persons, whether they are natural or constructive persons, depend for their existence and functioning on society. Without society—without social interaction, however tenuous—persons would not come into existence

[22] For more on this, see *infra*, pp. 33 ff.

in the first place and would not be able to survive and acquire those skills and characteristics that turn them from merely nascent entities into persons in the full sense of that term. Already Aristotle recognized this when he defined a person as a *zoon politikon*[23]—as a social being—where by this he did not mean that persons could not exist as solitary beings but that the realization of their potential to be persons rather than merely biological entities requires social interaction with other persons at some stage of their development. Therefore the very notion of personhood logically entails a context where the capacities of individuals as persons can be realized. This holds for all persons, which means that a society is necessary for all persons.

A society, in turn, is not an abstract entity that can exist independently of its members, nor is it an assembly of isolated individuals. It is a cooperative enterprise, and the constitutive members of such a cooperative enterprise are persons. Persons, in turn, have needs. These needs, however, are not only social: they are also material. This, in turn, means that in order to exist, the fabric of interaction among the members of a society has to be structured in such a way that both the material needs of its members as well as their psycho-social needs are met.

This, however—and this points to the fact that the Principle of Beneficence has a logical basis—is not possible if the actions of the members of society are entirely self-directed and for the benefit of the individual agents themselves.[24] Therefore when the needs of other members of society are at stake, then the thrust of the actions of individual members of society should not simply be for the sake of satisfying their own self-regarding needs but should also take the needs of these other members of society into account. Another way of putting this is to say that in order for society itself to exist, the framework of interaction among its members should be other-directed when this is appropriate, and hence should be beneficent as enunciated in the Principle of Beneficence. It goes almost without saying that, with due alteration of detail, this holds true both at the level of policy as well as at the level of individual actions.

However, the fact that the Principle of Beneficence imposes a general duty on all members of society as well as on society itself does not entail that the theoretical recipients of the relevant actions have an obligation to acquiesce in such actions. As was pointed out in the discussion of rights and duties above, rights and duties are correlatives. Therefore if there is a duty to act beneficently, then those who are owed that duty have a corresponding right. However, all other things being equal, whether they wish to exercise that right—and hence to activate the duty—is a matter of choice on their part, which in turn is a matter of their values. Consequently they may forego that right. More particular to the present context, this means that—once more, all other things being equal—they may refuse to receive the health care that would otherwise be their right if they decided to do so on the basis of their values. As to the qualifying clause "all other things being equal," this will be more fully in

[23] Aristotle, *Nicomachean Ethics*, Book I,2 (1094a-b).

[24] For a somewhat different analysis but to the same effect, see Rachels J. "Ethical Egoism." In *Reason & Responsibility: Readings in Some Basic Problems of Philosophy*, edited by Joel Feinberg and Russ Shafer-Landau, 532–40.

the subsequent chapters under the rubric of informed consent. For now, suffice it to say that if that refusal is not competently made, then it will not be valid; and if that refusal threatens the health, welfare or even existence of other member of society, then it will also not be valid because it engages the Principle of Impossibility as well as the Principle of Nonmaleficence.

Furthermore, what constitutes "the good of others," as this was enunciated in the preceding formulation of the Principle of Beneficence, also requires comment because it entrains considerations that are grounded in the Principle of Autonomy. More specifically, what constitutes the good is not an objective and absolute matter. It is a function of the values framework that is held by the individual who makes such a judgement. Therefore the question arises is whether—and if so, how—this is reflected in the Principle of Beneficence.

The short answer is that the Principle as it was enunciated above was stated in incomplete form. When stated in its complete form, it reads as follows:

> Everyone has a duty to advance the good of others insofar as they can do so without undue risk to themselves, where the nature of the good is in keeping with the ethically valid values of the recipients of the action in question.

This, of course, may make it occasionally difficult to put the Principle of Beneficence into practice since different persons may have different values, and one should never assume that their own values are the only ethically valid values.

But it does more than that: It immediately raises the question of what constitute ethically valid values. Nor is this a small matter because not all values are ethically valid. To recur to what was said in Chap. 1, values are action potentials[25]: They determine whether a right-holder will exercise a right that they have and they also determine whether a duty-holder will act in accordance with the duty that they have.

However, values have a certain content or nature, and that is the issue of ethical validity. For, not all values are ethically valid. For instance, values that do not consider children, the mentally disabled or individuals who have certain gender preferences as full and equal persons are not ethically valid values, as neither are values that discriminate on the basis of such things as skin colour, religious belief or ethnic origin. Values are ethically valid if and only if they respect persons insofar as they are persons and do not violate any of the fundamental ethical principles that structure the overall framework of ethics. Consequently, if the value that motivates a given action violates the Principle of Autonomy on part of another person where that other person holds to otherwise ethically defensible values, then that value will not itself be an ethically valid value.

With this, the justification of the limiting clause in the second formulation of the Principle of Beneficence becomes patent: Values that are inconsistent with the logic of autonomy and equality fail to fit the overall logic of the overall field of ethics in which rights are situated. Another way of putting this is to say that if the values that are used for any action are logically inconsistent with what it is to be a person—for instance, if they treat a person simply as a material object or as a production unit,

---

[25] Chapter 1, p. 2.

as happens in triage contexts when senescent individuals are considered to have a reduced right to health care (or indeed no right at all) on the basis of such measures as incremental cost-effectiveness ratios[26]—then such values are inconsistent with the Principles of Autonomy and of Equality as these were outlined above.

But, once more, that is not the end of the matter. The framework of ethics is not like the framework of games. One can construct games that can violate the Principle of Impossibility not only with respect to the laws of logic—witchcraft games that violate the law of the excluded middle are a case in point—but also with respect to the laws of nature, as for instance games that involve time travel or telekinesis. The framework of ethics, since it is a framework that governs actions, must be logically consistent with the Principle of Impossibility as it applies in the framework in which the matter is at issue. The framework of ethics is the framework of actions or situations in real life. Therefore it is a framework of actions that also has to be in keeping with the facts of reality. However, not all values that are logically consistent or compatible with ethical principles are compatible with the facts of reality. For instance, and more particular to the context of health care, the value of eternal physical life for human beings poses no logical issues. Thus, one can consistently conceive of a possible world in which human beings live forever. However, this is not the world in which we live, and it presents as possible something that is impossible given the physical and physiological facts of the world in which we live. More specifically, it is inconsistent with the second law of thermodynamics because only isolated physical systems can exist forever, and human bodies are not isolated systems. It is also inconsistent with the laws of physiology as they apply to the cells that make up human bodies because these cells have a limited life-span before they have to be replaced, and the Hayflick factor limits the number of times that human cells can replace themselves.[27] Competent values, therefore, have to be compatible not only with the Principles of Autonomy and Equality, but also with the Principle of Impossibility in the full sense of that term. This limits the array of values that ethically may be used when implementing the Principle of Beneficence.

2.2.6 *Principle of Nonmaleficence*

Closely related to the Principle of Beneficence—in a sense, its mirror image—is the Principle of Nonmaleficence

> Everyone has a duty to prevent harm to others insofar as this is possible without undue risk to themselves.

However, as should be clear from the preceding discussion of the Principle of Beneficence, simply formulating the principle like this is to ignore the fact that just as what counts as good is relative, so is what counts as harm. Both are functionally

[26] Cf. Appleby et al. (2007).

[27] Hayflick and Moorhead (1961).

determined by values. Consequently for an agent simply to assume that the notion of harm is a universal notion or that the agent's own values regarding what counts as harm should determine how the Principle of Nonmaleficence should be implemented would be to ignore that the recipient of the action is a person and hence may have their own values, and that these may not be the values that are espoused by the agent.

So, for instance, not providing a patient with pain medication would be considered a violation of the Principle of Nonmaleficence if the nature of what counts as harm were universal and accepted by everyone. On that assumption, pain is considered an evil and not to assuage it if it was possible to do so would be to cause harm. However, providing effective pain-reducing medication may cause confusion or have other side-effects for the patient, where according to the patient's own values these side-effects would themselves fall into the category of harm. Therefore while the patient might well agree that pain is a harm, according to the patient's own values[28] the confusion and other side-effects that would result from the administration of narcotic analgesics or similar pain-reducing medication might be rated as more serious and less desirable than the alleviation of the pain. Nor would such values automatically count as incompetent simply because they would not be common or would simply be different from the standard perception of pain.[29] Consequently a patient may reject a pain mitigating intervention even though in the physician's mind and according to standard medical values harm would result from acceding to the patient's ultimate decision.

Or, to consider another example, strictly speaking turning off the ventilator of someone who is ventilator dependent or administering a pharmacological agent that will painlessly kill a patient is to engage in homicide because it is to bring about the death of the patient. In general, both in ethics and in law, homicide—which is to say, ending a person's life—without sufficient excuse is considered to be a harmful act. However—and this is where the notion of sufficient excuse comes in and where the competent values of the recipients of such action become relevant—a patient may find their continued existence itself and, more specifically, the manner of their continued existence, to be a violation of their values. For instance, they may find that being kept alive when their level of pain or nausea cannot be controlled,[30] or when they are completely dependent on other persons for all of their bodily functions and are mentally so impaired that they can no longer form cognitively meaningful memories and can no longer recognize other individuals to be a violation of their dignity as persons and to be a violation their values. Such values are ethically defensible because, religious and social constructs aside, they are grounded in the Principle of Autonomy, which applies to them insofar as they are persons. That does not mean that all persons should share such values. It merely means that they are ethically

[28] Lewis CS. *The Problem of Pain*, available at https://www.google.ca/books/edition/THE_PROBLEM_OF_PAIN/cXgrDAAAQBAJ?hl=en&gbpv=1&printsec=frontcover; Yancey and Brand (1997).

[29] Benedetti et al. (2013), Brock et al. (2013). It bears emphasizing this this perspective on pain is distinct from suffering from algolagnia.

[30] Intractable pain falls into this category. Cf. ICD-10 R62. Cf. Viswanathan and Burchiel (2013).

defensible values. Consequently for third parties to engage in what otherwise would be ethically appropriate actions by keeping the individuals who held these values alive would actually be to violate the competent values of the individuals who were being kept alive. It would be to cause harm, and would therefore be a violation of the Principle of Nonmaleficence. As an aside, these considerations are even reflected in the World Medical Association's injunction that an ethical physician "shall respect a competent patient's right to accept or refuse treatment" and "respect the rights and preferences of patients."[31]

Taking these considerations into account, therefore, the initial formulation of the Principle of Nonmaleficence that was given above has to be augmented by the addition of a limiting clause as follows:

> Everyone has a duty to prevent harm to others insofar as this is possible without undue risk to themselves, where the nature of the harm is in keeping with the competent values of the recipients of the action in question.

2.2.7 *Best Practices*

The distinction between providing care and providing the best possible care that was mentioned a little while ago raises an issue that sometimes surfaces in discussions of quality management[32] systems under the heading of best practices. It is generally assumed that if there is a duty, then there is an overarching duty to fulfil that duty in the best way possible. The assumption can be stated in the form of a principle:

> Whoever has a duty has a corresponding duty to try to fulfil that obligation in the best way possible.[33]

The principle itself is assumed to be grounded—if an attempt is made to ground it at all—in the logic of what it is to be a duty. The argument here tends to take the following form[34]: A duty is an obligation to do or to refrain from doing something. Therefore, an obligation is not fulfilled if the relevant action is not carried out at all, if it is carried out only partially, or if the obligation to refrain from doing something is not met at all or it is only partially adhered to. Being part of what structures the framework of ethics; this applies to all duties, and therefore applies without exception. Consequently—so it is argued—while the supposed principle does not ground the

[31] World Medical Association, *International Code of Medical Ethics;* accessed 09/06/2020 at https://www.wma.net/policies-post/wma-international-code-of-medical-ethics/.

[32] Examples of this are standards that are promulgated by the International Standards Organizations under ISO 9000 and ISO 1400.

[33] Where an unstated but understood condition is that, one cannot have an unethical obligation. This condition becomes relevant when distinguishing between ethically valid and purely contractual obligations.

[34] CF. International Medical Informatics Association, Code of Ethics for Health Information Professionals, Principle 6; accessed 11/06/2020 at https://imia-medinfo.org/wp/wp-content/uploads/2019/08/IMIA-Code-of-Ethics-2016.pdf.

existence of conceptually distinct duties, it supposedly in general conditions whether a duty is satisfied.

However, closer consideration of this reasoning shows that it is not quite trenchant. If it is supposed to show that this holds without exception, then it confuses what is stipulated by a duty—its substance, so to speak—with something that is supererogatory. That is to say, a duty stipulates what should or should not be done. It thus stipulates the conditions that must be met in order for the duty to be fulfilled. This in turn means that if these conditions are met then the duty is satisfied, whereas when they are not met then the duty is not satisfied.

The substance of a duty, however, is neutral as to the manner of its satisfaction and it is silent about what falls outside of the conditions that it stipulates. Best practices, however, focus on the manner of satisfying a duty. They indicate a way of fulfilling that duty and, in particular, what is considered to be the best possible way of doing so. Differently, best practices take account of various possible ways of satisfying a duty and from among these choose what are considered to be the best possible way of doing so. Therefore they go beyond what is necessary to fulfil a particular obligation. That is why they are supererogatory.

Logically speaking, therefore, while it may be praiseworthy to follow best practice guidelines when faced with a particular duty—for instance, when faced with the duty to provide emergency care for a person who suffers from a cerebrovascular accident—and while it may reduce the likelihood of successful litigation, the substance of the duty to provide such care itself is satisfied if appropriate measures are implemented even though these may not be the best possible measures.

Moreover—and this addresses the claim that the reasoning in support of best practices is not quite trenchant—to leave matters here would be to underplay the relevance of the context when evaluating whether a duty is fulfilled. As was pointed out in the discussion of the Principle of Impossibility, one cannot have a duty to do what is impossible, and that this is conditioned both by logic and by the context in which the corresponding right is claimed. As was also pointed out, this context is not only logical in nature but also includes the material resources that are available and the equal and competing rights of others. Therefore fulfilling a duty in the best way possible may require resources that are necessary to meet otherwise legitimate obligations. An instance of this would be providing the best possible treatment for a particular health condition, but that this would be so expensive as to reduce the amount of resources that would be available for providing life-saving interventions such as cardiac by-pass surgery or acute appendectomies.

Consequently, if there are different ways of dealing with a particular health condition and if they are equally effective, then the Principle of Nonmaleficence entails that one should choose the one that would involve the least drain on the resources that are necessary to meet the health care needs of other persons. If following best practices is understood in this latter sense, then doing so is ethically appropriate; however, if it is not understood in this manner, then following best practices would be supererogatory and may not be ethically valid. This distinction between fulfilling a duty and engaging in a supererogatory act is important in the delivery of health care, which is always context- and resource-dependent. Therefore it becomes important when dealing with the ethics of resource allocation at the policy level, which will be discussed in Chap. 3.

2.3 Conclusion

The aim of this chapter has been to indicate how the Principles of deontological ethics are logically related to the concept of a person, and how they form an integrated framework that determines how ethical rights and duties devolve from them. The analysis, therefore, has proceeded independently of legal, social and psychological considerations. If what has thus been presented is correct, then the Principles themselves are objectively valid just as soon as there are persons and these persons exist in a social context—which, as has also been suggested, is integral to being a person rather than being a merely biological entity.

An important implication of this, then, is that the Principles and the ethical framework that they structure hold independently of legal, societal or even individual psychological recognition—in other words, that they have objective validity independently of whether they are accepted as such—and that they have atemporal validity even when historically this has not always been recognized. A further implication is that they have also global validity, since they hold as soon as there are persons. Finally, the implication of what has been presented in the body of this analysis is that the rights and obligations to which they give rise hold for all persons irrespective of whether they are natural persons or corporate (constructive) in nature—something that becomes important in the next chapters when considering whether there is a right to health care and what the corresponding duties of such a right may be for the various parties. The fundamental question, of course, is whether there is such a right. The next chapter is devoted to dealing with this issue.

References

Appleby, J., Devlin, N., Parkin, D. 2007. NICE's cost effectiveness threshold. *BMJ* 335 (7616): 358–9.

Benedetti, F., W. Thoen, C. Blanchardc, S. Vighetti, and C. Arduino. 2013. Pain as a reward: Changing the meaning of pain from negative to positive co-activates opioid and cannabinoid systems. *Pain* 154 (3): 361–367.

Brock, B., J. Jetten, M.J. Hornsey, and S. Laknes. 2013. *Personality and Social Psychology Review* 18 (3): 256–279.

Chafee, Z, Jr. 1919. Freedom of Speech in War Time, 32 *Harvard Law Review* 932: 957.

M. Chen-Wishart, A. Loke and B. Ong (eds.). 2016. *Studies in the Contract Law of Asia*, Oxford U Press, 373: 391.

Hayflick, L., and P.S. Moorhead. 1961. The serial cultivation of human diploid cell strains. *Experimental Cell Research* 25 (3): 585–621.

Reuter, P. 1985. *Introduction au droit de traités*, 152–157. Paris.

Sadikov, O.N. 2015. Soviet Civil Law. In *The Law of Obligation*. Routledge.

Tannsjo, U.T. 1999. *Coercive Care: The Ethics of Choice in Health and Medicine.* London: Routledge.

Viswanathan, A., and K.J. Burchiel. 2013. Intractable pain. *Neurosurg Focus* 35 (3).

Yancey, P., and P. Brand. 1997. *The Gift of Pain.* Zondervan, Publishing/Harper Collins.

Chapter 3
The Right to Health Care

Abstract The present chapters argues that there is a right to health care, that this right is grounded in the fundamental principles of ethics, and that an ethical society has an obligation to supply such care insofar as it is able to do so within the context of its other obligations and the limits of its resources. It also distinguishes between health care services construed as commodities as opposed to rights, and outlines some tools that may be used in making allocation decisions at the policy level.

Keywords Allocation · Commodity · Research · Age-based rationing · Palliative care

3.1 Introduction

The analysis in the preceding chapters has served two purposes: to present and justify the ethical approach that forms the basis of this study and to clarify the terms that will be used in the analysis of the question whether there is a right to health care. The purpose of the present chapter is to argue that there indeed is such a right. Further, it will be an underlying assumption in this chapter as well as in the chapters that follow that there is a fundamental difference between the right to health and the right to health care. As was pointed out in the Preface, there is not and cannot be a right to health because that assumes that in principle it is possible to guarantee that people will be healthy, which is not the case. There may be situations—incurable conditions are a case in point—where the very nature of the individual guarantees that the individual will never be healthy. If there was a right to health that would mean that a society would always be guilty in not ensuring that the individual was healthy.

However, the right to health care is another matter. It is the right to interventions whose purpose is to ameliorate health conditions that interfere with an individual's ability to take equitable advantage of the opportunities that exist within a given society. Such care may not put an individual into a complete state of health, but for all that this would be ethically acceptable. As the United Nations Committee on Economic, Social and Cultural Rights stated in its "General Comment No. 14", there

E.-H. W. Kluge, *The Right to Health Care: Ethical Considerations*, The International Library of Bioethics 92, https://doi.org/10.1007/978-3-030-93838-3_3

is a fundamental right to "the highest attainable standard of health",[1] and bases this on Article 25 of *The Universal Declaration of Human Rights*. What counts as the highest attainable standard of health is functionally related not only to the specific condition of a particular individual and the prevailing conditions in a given society but also to the resources that the society has available for providing health care: and this is the subject of the present chapter.

Focusing on the issue of social resources—the rights of individuals will be addressed more specifically in Chaps. 6, 7, and 8—all rights exist in a soil context, and the right to health care is no exception. On the one hand there are the right-holders, and on the other hand there are those who have the corresponding obligations. Societies, however, are always limited in nature. There are no infinitely large societies. This means that the number of people who can respond to a right-claim—and in particular to the claim for health care services—will always be limited. Consequently some way must be found to resolve in an ethically acceptable fashion the competition among various claims to health care when not all individuals can be served at the same time. Furthermore, societies also only have access to a limited amount of resources, because a finite number of people can only produce a finite amount of goods. This complicates the allocation question even further.

Moreover, if the analysis of the right to health care that is offered in this chapter is going to be ethically grounded but also usable, it must not only satisfy deontological ethical principles but must also provide a way to harmonise the considerations that arise at the policy level of planning—which determine the structure of the overall framework—with considerations that arise at the hands-on level at which health care is actually delivered. An attempt will be made to show how this can be done by introducing the notion of an act-effectiveness coefficient and by indicating how it would function in actual practice.

3.2 Preliminary Remarks

3.2.1 Rights Versus Commodity Orientation

Before beginning with the analysis itself, however, it may be appropriate to clear up a possible misunderstanding that may be potentiated by the way the question of whether there is a right to health care is sometimes posed.

That is to say, there are two fundamentally distinct views on the nature of health care itself: one construes it as a right, and another construes it as a commodity.[2] The argument then proceeds by saying that these two views lead to distinct approaches to the delivery of health care. The rights-oriented view uses ethical principles to ground the right to health care and to structure an appropriate health care service

[1] https://undocs.org/E/C.12/2000/4.

[2] Cf. Aggarwal et al. (2010), Petrou and Vandoros (2018).

delivery system. In other words, it does not deal with health care services on a purely economic basis but considers them in ethical terms and decides on an ethical basis which services will be offered, how they will be offered and how they will be distributed.

By contrast—so the argument continues—as the phrase itself indicates, the commodity-oriented view of health care treats health care services like commodities. Therefore whoever cannot pay for them will not get them, and the issue of ethical rights does not enter the picture at that level. Consequently, a commodity-oriented perspective uses economic calculations to determine the nature, range and distribution of the health care services that are being offered—and indeed whether they will be offered at all—and it will appeal to ethical principles only insofar as these determine the limits within which all economic activity has to be conducted.

3.2.1.1 Distinct Views of Society

However, to approach the question of whether there is a right to health care in this fashion is to approach it in a fundamentally skewed manner because it equates the right to health care with the services that are used in providing that care. To put it bluntly, both what have been called the rights model and the commodity model agree that health care services have to be paid for, which means that both models understand that health care services are commodities. Where they differ is in how they construe the notion of society itself and consequently in how those commodities should be provided.

That is to say, the rights-oriented model is based on a perspective that views society as a person—to be sure, as a constructive person as this notion was outlined in Chap. 1, but as a person nevertheless—and therefore as being subject to the same ethical principles as apply to all persons. On that basis it argues that the Principles of Ethics—and in particular the Principle of Equality—entail that society is obligated to provide health care for its members on an equitable basis.

By contrast, the commodity-oriented model does not construe society in that fashion. It sees society not as a person but as a collection of individual who have social, economic and other interests. In other words, it sees society as an aggregate of people who live together in a more or less ordered community with more or less shared traditions, institutions, and interests. Therefore, while it may agree that ethical considerations apply to the members of society, it does not consider society itself as being subject to the Principles of Ethics because society is not a person. Ultimately, therefore, it sees economic and legal parameters to be the determinants of how the overall social framework is or should be structured.

3.2.1.2 Some Calculative Tools: Ethical Considerations

This way of putting the issue not only clarifies what is actually at stake, it also shows that certain calculative tools that are sometimes used to determine the appropriateness of health care interventions are used inappropriately when they are used in a rights-oriented approach. Such tools include Quality-Adjusted Life Years (QALY), Disability-Adjusted Life Years (DALY)[3] and Health-Adjusted Life Years (HALY).[4] QALY is a measure of the burden that a particular disease or health condition would place on whoever has the disease or condition in order to assess the value of engaging in medical interventions. It combines measures of the expected quality of life and the length of time that the sufferer could reasonably be expected to live with the nature of the intervention.[5] DALY is a measure of the overall burden that a disease or health condition presents by calculating the number of years that are lost due to ill-health, disability resulting from the condition and the [6]number of years of projected life. HALY involves similar calculations except that it focuses on overall health, not simply a specific disability or condition.

However, while these and similar measures may be usable when adopting what has been called a commodity approach to health care, their use presents a logical problem for a rights oriented perspective because they contain a hidden assumption that is inconsistent with the logic of that perspective: the assumption, namely, that that the strength of a person's right to health care is a function of the time that they can reasonably be expected to live and of the quality of such lives.

Before expanding on this, it may be useful to return for a moment to the fact that a rights-oriented perspective on the nature of health care must also treat health care services as services, which is to say as amenities that have an economic value and therefore also have a price. As was pointed out above, every society is finite and hence has only a finite amount of resources. This means that even if health care is construed as a right, the question of how to structure the framework that provides the actualisation of that right—in other words, that actualises that right by providing appropriate services—has to take economic considerations into account because it has to pay for these services, and it must ultimately face the question of resource allocation both at the level of individual access to health care services as well as at the policy level of what types of services to provide. Consequently even a rights-oriented perspective on health care ultimately has to face the question of who shall have when not all can have; and the only way that this question can be answered is to apply some way of adjudicating between competing rights claims.

However—as will be discussed more fully later—since a rights-oriented perspective on health care is based on the assumption that health is a major determinant

[3] World Health Organization, WHO methods and data sources for global burden of disease estimates 2000–2011; accessed 19/06/2020 at https://www.who.int/healthinfo/statistics/GlobalDALYmethods_2000_2011.pdf.

[4] Gold et al. (2002).

[5] La Puma and Lawlor (2009), Weinstein et al. (2009).

[6] Arnesen (2011).

of a person's ability to take advantage of the opportunities that are available within society, it argues that if a society wants to be just and equitable—in a word, if it seeks to align its policies and structures with the principles of ethics that were outlined in the preceding chapter—then it will have to attempt to create a level playing field for all its members. Therefore such a society will have to try to minimize any health-based differences among its members if these differences would interfere with their ability to take advantage in an equitable manner of the opportunities that are available within that society. Consequently, such a society will have to attempt to create a health care system that will meet the legitimate[7] health care needs of its members in order to minimize the effect of such health-based interpersonal differences, and to balance competing rights.

Of course this could be done in a variety of ways. For instance, it could be achieved by providing each of its members with sufficient resources to purchase the health care services that are provided by private health care suppliers who offer their services on a commodity fee-for-service basis.[8] Alternatively, while still recognizing that health care services are commodities, it could achieve its aim by providing appropriate funding directly to private and commodity-oriented health care service providers so that the individuals who require their services would receive these services as and when they are required them but without the individuals themselves having to defray the costs. Still differently, a society could actualize a rights-oriented framework by setting up publicly funded not-for-profit facilities that could be accessed without charge by any member of society in order to deal with their health care needs. Further variations on these options would include requiring citizens to pay according to their ability, to pay nothing at all or to pay in hidden form through methods of taxation. Still other options are possible, but the general thrust of acceptable options should be clear.

The problem is that no matter what option an ethically oriented society adopts, it has to deal with the issue of allocation at three levels: at the level of societal resources, the level of overall health care, and at the level of individual cases. At the level of societal resources it has to deal with the question of how to allocate sufficient resources from its overall resource pool in order to defray the costs that are associated with health care, because the overall resource pool still has to be divided among other legitimate social expenses such as education and infrastructures. At the level of overall health care, it has to determine what types of health care services should be covered by the framework in general; and at the level of individual cases it has to determine which individual shall have access to services when individuals compete for the same service and that service is in limited supply.

At first glance measures such as QALY DALY and HALY would seem to be useful tools for dealing with these issues because they would provide comparative

[7] The reason for this qualifier will become more apparent in the subsequent discussions of auto- vs. hetero-induced health care needs.

[8] There are various ways in which this could be achieved: E.g., by giving tax credits for insurance, by giving direct subsidy, by providing employers with credits so that they could insure their employees, etc.

measures for making allocation decisions both as to types of services in general as well as with respect to the services that should be provided on an individual person-oriented basis. Moreover, they seem to have no inherent economic parameters, which is why it would seem appropriate to use them even from a rights-oriented perspective of health care. It would only be when economic parameters are attached to them that they would be inconsistent with the ethical underpinnings of a rights perspective, because only then would the right to health care and its strength be determined on the basis of cost.

However—and this brings the discussion back to the claim that was mentioned above, namely that there is a hidden assumption in using such measures that was mentioned above—although the hidden assumption is not inherently economic in nature, it nevertheless is inconsistent with the logic of a rights-oriented approach to health care. For—and this was also mentioned above—to use such tools is to assume that the existence and strength of a right to health care, whether that be on an individual basis or at the policy level, is functionally determined by the length and the quality of that life. And that is not the case.

To put the reason for saying this first in general terms regarding rights themselves, the strength of a right is functionally determined by the ethical principles that structure the framework in which the right arises, the condition under which it arises, and how it relates to the other right-claims that arise under this framework. The Principles of Autonomy, Equality and Impossibility are here implicated: The individual must ethically be justified in claiming the right, such a claim must not abrogate the equal status of other right claimants, and it must be possible for the corresponding duty-holder to fulfil the claim under the circumstances that obtain.

Applying this to the issue of the right to health care—assuming for the moment that there is such a right—if the right to health care is invoked by members of society, then fulfilling that right must not make it impossible for society to meet its other social obligations. This would be the problem of determining the size of the overall health care budget in competition with other legitimate social duties. The next issue would then be to decide what types of health care services to fund from the available budget of health care itself. This would be the problem of macro-allocation within the overall health care framework. Finally, there would be the problem of micro-allocation: of how to allocate the limited resources that are available within each type of service that had been decided upon in terms of macro-allocation in order to satisfy the individual claims to such a particular kind of service when these are made on a person-to-person basis.

As was said before, using measures such as QUALY, DALY or HALY to evaluate the existence and strength of a right to health care would seem to fill the bill because these measures do not directly involve economic considerations, and therefore would seem to be acceptable tools under a rights perspective. However, as was said before, they provide a ranking of health care rights that is based on the differences between lives as these are calculated on the basis of the length of the lives and of their functional characteristics. Therefore, logically speaking, they effectively are measures of the health-based advantages or disadvantages that are faced by the individuals who would lead such lives. Logically speaking, however, health-based advantages and disadvantages are measures of health-based differences in need-status of individuals and they determine whether the individuals can take advantage in an equitable fashion of the opportunities that are afforded by society to its members. By that very token, and independently of whether these measures are used at the individual- or the policy-level, using them to determine the existence and strength of a given right to health care would be to use measures that discriminate on the basis of differences in the amount of health care needed.

However, as was said, differences in health care need ground differences in the ability to take advantage of the opportunities that are available in society. Therefore using such tools would effectively be to legitimate discrimination on the basis of differences in the ability to take advantage of the opportunities that society offers to its members. By that very token, it would undercut the basic ground that would justify the rights approach to health care itself—namely, the Principle of Equality. In that way—as will be argued more fully in a moment—it would stand in direct conflict with what justifies a rights-oriented approach to health care in the first place.[9]

This is not to deny that such tools as QALY, DALY and HALY could be used in a rights oriented approach to health care. As was said a moment ago these tools—and others like them—can provide a clear picture of the functional advantages and disadvantages of health conditions and of the effect that treatment or lack of treatment would have. Therefore using them would allow the right-holders[10] to decide whether they would wish to exercise their right to health care, and therefore they may well be important parameters when they consider whether to exercise their right to health care on a given occasion. In other words, they would be legitimate factors under the Principle of Autonomy. They would not, however, be legitimate factors for a society or a health care institution to consider when deciding whether there is a right to health care in the first place or what the strength of that right would be.

3.3 The Right to Health Care: The Argument

3.3.1 Society, Personhood and Related Matters

Turning, then, to the initial question of whether there is a right to health care, the answer may be found by going back to what was said about how the nature of society.

As was then said, there are two distinct views of the nature of society: the first sees society as a functional complex of individuals who form a more or less integrated framework. It is grounded in a view of personhood that construes persons as entities that have the capacity for cognitive awareness and volition, and that considers that there are wo sorts of persons: natural persons on the one hand, and constructive

[9] Cf. O'Neill (2001) at 15: "Nobody can accept two or more principles or rules whose enactment is intrinsically incompatible.

[10] The reason for stating the issue in this fashion will become clear when dealing with the issue of macro-allocation.

persons on the other. And while it acknowledges that natural persons are distinct as biological entities, it also understands them to be dependant not only for their existence but also for their very nature *as persons* on being embedded in a social context. The other view sees society as a collection of distinct individuals who associate more or less closely with each other but who can exist independently as rational beings of any such association.[11] These two distinct views of society and personhood may be called a *fusional* and a *clustering* view of society respectively; and while a fusional view construes society itself as a person—albeit as a constructive person—a clustering view of society does not.

Not surprisingly, it is these two distinct views of society that lead to fundamentally distinct views on the right to health care: The fusional view construes health care as a right because, simply put, as constructive person, it has an ethical duty to structure its framework in accordance with ethical principles, which means that it has a duty to structure it in a way that provides its members with the possibility of taking advantage in an equitable manner of the opportunities that it offers to its members since the health status of individuals is a primary determinant of their ability to do so. By contrast, a clustering view does not construe health care as a right but as on a par with other commodities because a clustering view does not consider societies to be persons. Therefore the Principles of Ethics that were outlined in Chap. 2 do not apply to it, but only pragmatic concerns are of relevance.

More specifically still, and referring back to what was said in Chap. 2, a fusional view of society, being grounded in an understanding of biological persons as social animals in the Aristotelian sense of *zoon politikon*—i.e., as biological beings who can actualize their capacities as persons only through interactions with other persons—construes society as providing the framework within which such actualizing occurs.[12] Since, as was also pointed out in Chap. 2, societies in this sense are subject to ethical principles, it means that the fusional framework must integrate these principles into its structure if it wishes to be ethical because it is a constructive person. Among other things, therefore, it brings the Principle of Equality into play; which in turn means that its social framework should ensure that all members of society are treated equitably. If it does not do so then the society, *qua* constructive person, would violate the Principle of Equality in its actions.

As was pointed out before, this is not the case with the clustering view of society (which is associated with an atomic view of personhood). While a clustering society may in fact structure the framework of interaction among its members according to ethical principles, it need not do so because, not being a person, it is not subject to these principles. As long as interactions among its members are in line with the interests of its members and do not reduce to chaos, no further requirements exist;

---

[11] Arguably, both Buddhist and Hindu views of personhood constitute two further views of personhood: the former seeing true personhood as anatman (being a non-person), the latter seeing personhood as centering in atman (being a true self). However, logically speaking, both views are subspecies of the functional view of personhood and differ merely in the emphasis that should be placed on individual and focused perceptions and interrelationships.

[12] For a contemporary study from a non-philosophical point of view, see Cacioppo and Patrick (2008).

and this, in turn may be achieved through a political process that plays off the interest of its members against each other. Adherence to ethical principles, therefore, is not a necessary feature of this type of society unless it can be shown that adherence to ethical principles was necessary to avoid a chaotic system, and unless it could be shown to be in the interest of the preponderance of its members. And even then, it would not be a feature of this type of society unless an ethical process that governed social interactions was to put in place. That, however, would be an ad hoc intrusion of ethics into the process, and would not be inherent in the nature of a clustering view of society itself.

Before going any further it may be appropriate to make explicit a distinction that was mentioned in Chap. 2 when discussing the Principle of Equality but which, in view of the subsequent analysis, deserves separate mention. It is the distinction between treating persons the same way and treating them equitably. To treat persons in the same way is to ignore any ethically relevant differences that may exist between them; by contrast, to treat them equitably is to acknowledge any ethically relevant differences and to accommodate them by taking appropriate steps to eradicate or minimize their effect on the individuals who exhibit them.

3.3.2 *Some Consequences*[13]

3.3.2.1 The Influence of the Principle of Equality

Returning, then, to the question whether there is right to health care, it seems reasonable to assume that because health is a major determinant of a person's ability to take advantage of the opportunities that are available within a society, a fusional view of society that adheres to the Principle of Equality will attempt to create a level playing field for all of its members and try to minimize health-based personal differences by treating health care as a right insofar as any health-based differences between its members would affect their ability to take advantage in an equitable manner of the opportunities that are available in that society. As was said before, logically speaking, this could be achieved in several different ways. However, they would all have to respect the Principle of Equality in the sense that was just outlined.

3.3.2.2 Curtailment of Autonomy

However, the Principle of Equality cannot be the only ethical determinant, since a fusional society could not ethically be selective in its adherence to ethical principles but would attempt to be ethical in the full-blooded sense of that term. Among other things, this means that such a society would also accord full and appropriate weight to Principle of Autonomy. Therefore all other things being equal, such as society

[13] In what follows, the term "society" will generally be understood as referring to a fusional society.

would acknowledge that the actual access to health care services would usually be on a voluntary basis since otherwise health care would become a duty and cease to be a right.

At first glance, however, this suggests that if such a society adhered to the Principle of Autonomy, it would not be able to mandate that its members seek appropriate health care even when the individuals had a condition that was communicable and which, if left untreated, would seriously threaten the health of other persons but the affected individuals decided not to do anything about it. That, however, seems to mean that, under such circumstances, for an ethical society to structure its framework in accordance with the Principle of Autonomy would lead to a head-on collision between the Principle of Autonomy and the Principle of Nonmaleficence, and therefore that the principles of ethics would not be logically consistent after all, as it was claimed in the preceding chapter. For, as was just said, there are health conditions—serious communicable diseases such as Ebola or COVID-19 are good examples—where, if a society does not turn the right to health care into a duty relative to the sort of condition in question, harm to other members of society may result and the very existence of society itself might be threatened.

However, to argue thus would be to paint a skewed picture. It does not mean that the Principles are in collision, but that the limiting clause of the Principle of Autonomy—"subject to the equal and competing rights of others"[14]—then comes into play, and that society has a duty to structure the framework of interactions in accordance with this limitation. In other words, it means that the relevant individuals would be ethically obliged to minimize the chance of transmission by exercising their autonomy in keeping with this limitation and by engaging in appropriate and effective social distancing or by isolating themselves so that the wellbeing of other persons would not be violated, etc. It is only when this was not reasonably possible—for instance when transmission could occur despite social distancing or solation—or if it was reasonable to assume that the affected individuals would not actually take the appropriate steps, that the society would be ethically obliged to put in place appropriate administrative measures that would mandate the relevant harm-reducing actions on part of the affected individuals. In a sense, of course, this would effectively turn the right to health are into a duty. However, society's enactment of such measures would not be a rejection of the Principle of Autonomy but would more properly be described as being a formal and administrative recognition of the fact that the individuals who suffer from such conditions have a duty to limit their autonomy in keeping with the equal and competing rights of others, and that society has an obligation to ensure that the way it structures what constitutes acceptable behaviour among its members satisfies all aspects of the Principle of Autonomy itself.[15]

[14] Chapter 2, p. 23.

[15] For more on this, see Chap. 8, *infra*.

3.3.2.3 Autonomy, the Right to Health Care and Social Values

At the same time, it is important to note that the Principle of Autonomy also applies to a fusional society within the context of the limiting condition that governs autonomy, especially with respect health care when it comes to planning for and providing health care at the policy level. In other words, while an ethical society *qua* constructive person would also have the right to autonomy, this does not mean that it can choose not to provide health care and still be considered ethical.

To see why this is the case, it may be appropriate to go back to what was said in the preceding chapters about persons. As was then indicated and was emphasized above, the concept of person incudes not only biologically distinct individuals but also groups of individuals such as corporations, associations that function like persons and indeed societies themselves. Therefore, a society may also appeal to the Principle of Autonomy when deciding on its actions. Since autonomy in decision making is exercised on the basis of preferences or values,[16] this may be perceived to entail that a society—considered as a constructive person—may exercise the Principle of Autonomy and use societal values to decide whether—or how—to implement the right to health care. An instance of where this understanding shaped the reasoning of a society relative to the health care services that it offered to its members was the case of Oregon with its *Basic Health Services Act* of 1989.[17] After consulting with what it considered to be a representative set of its members in order to identify what could be considered social values, Oregon structured the health care services that it provided under its publicly funded Medicare and Medicaid programs in accordance with these values to include certain services and to rank these services in terms of priority. As was said, the underlying assumption here was that, being a constructive person,[18] society has a right to exercise autonomy when deciding on matters of health care and hence should be able to use social values when making its decisions, since these are the values of society itself.

However, issues of implementation aside[19] and ignoring the fact that the undertaking occurred within what in fact was a clustering rather than a fusional view of society that did not really espouse the view that health care should be seen as a right, the very supposition that underlay this attempt was basically flawed. While the Principle of Autonomy does determine whether a right is exercised[20]—and this of course includes the right to health care—and while societies when construed as constructive persons do have rights, it is not societies themselves that have a right to health care but its members. Therefore it is the members who have the right to exercise the right to health care, and it is their individual values that are relevant when it comes to exercising this right, and to what extent. This is not altered by the fact that,

[16] On the relationship between values and preferences, see Warren et al. (2011).

[17] Cf. Baker (1993). See also Nelson and Drought (1992).

[18] Cf. Chap. 1, p. 8.

[19] The *Act* underwent substantial changes from its original formulation for economic reasons, but did not revise the underlying value-based structure of 1993.

[20] For limiting conditions, see Chap. 2, pp, 22 f.

qua constructive person, a society may autonomously choose to exercise its rights. But, as was said, this does not apply to health care. It is not society *qua* constructive person that has a right to health care: It is its individual members. Society only has a duty. And while it may exercise autonomy in whether to adhere to that duty, not adhering to it would be an ethical default.

It is therefore a fundamental error to assume that social values may play any ethically defensible role in determining whether a fusional society[21] should offer health care services, what health care services it should offer, to whom they should be offered, or to what extent they should be offered.[22] This holds true even if it could be shown that the values of society are in some way derived from those of its members, because the derivation of the values does not change the nature of the right and it is the individual members who have that right, not groups of members.

3.3.2.4 The Right to Health Care and Economic Considerations

At the same time, it would be a mistake to assume that other ethical considerations drop entirely out of a picture so far as society itself is concerned. Whenever rights and duties are involved, the Principle of Impossibility is also engaged. It must not be impossible to fulfil the duty. After all, if it is not possible to satisfy the duty under the conditions that obtain and the duty holder is not responsible for voluntarily bringing about that impossibility, then the duty lapses. Therefore society can—and indeed must—satisfy its member's right to health care, but it must structure the framework of the health care services that it provides with reference to what is possible. This means that it cannot ignore outcome and cost/effectiveness considerations since the latter functionally influence the availability of resources and hence affect the possibility of providing the relevant services. Furthermore, population-based differences in health needs should also be factored into macro- decision making since such decisions must stand the test of equity.

However, even this comes with conditions, because how the Principle of Impossibility may be applied, and in what sense cost and outcome considerations are relevant, is not as simple as it may seem at first glance. To put the point initially in general terms, the fact that providing a certain kind of service would make it impossible to fund another and higher-ranked kind of service that was also mandated as a matter of obligation does not mean that therefore the duty to provide the first kind of service is always rendered ineffective. That would follow if and only if there were no alternative ways of satisfying the right that was associated with the first kind of service.

However, it may be the case that this right has several ways of being satisfied. By way of analogy, if there is a right to education and society has a corresponding duty

[21] In what follows, and unless otherwise stated, "society" will be understood in the sense of fusional society.

[22] The author was in error in this regard in the article "Social Values, Socioeconomic Resources and Effectiveness Coefficients: An Ethical Model for Statistically-based Resource Allocation," in R Cohen-Almagor (2000).

to provide educational services to its members, then this duty could be satisfied in several ways: It could be satisfied by paying for private tutoring for each child, by instituting a public education system to which all children have access, by putting in place an electronically mediated teaching system and supplying the necessary means for accessing the system, and so on. However, the fact that paying for private tutoring would be so expensive that it would make it impossible for society to defray the costs of providing a workable infrastructure or even of providing for health care would not mean that therefore the duty to provide educational services would lapse. It would merely mean that society would have to explore the other available options.

Furthermore, not all right and duties have the same priority ranking. For instance—to continue with the example that was just considered—the right to an education is lower in priority than the right to life, because one cannot receive an education if one is not alive. In fact, logically speaking, being alive—existing as a person—is a precondition for having any rights at all. Therefore in situations where satisfying one right makes it impossible to satisfy another, the priority status of the competing rights also has to be considered.

Applying these considerations to the question of what role the Principle of Impossibility plays in implementing the right to health care, it means that if the right to a particular type of treatment for a particular type of condition is trumped by the higher-ranking right of treatment for another condition, then this does not necessarily mean that the right to health care services for the former condition lapses. The point here to keep in mind is that the right to the treatment for a specific health condition is merely an instance of the general right to health care itself, and that the right to health care as it is engaged by a particular condition is not modality-specific. Furthermore, as was pointed out in Chap. 2, while the existence of a given right entails a corresponding duty, that does not mean that duty will be met only if the right is fulfilled in the best way possible.[23] The duty will be met as soon as and as long as the conditions that are specified in the right itself are satisfied. Anything more would be supererogatory.

Therefore, when considering the role that the Principle of Impossibility plays in making health care decisions at the policy or macro-level, it is important to keep in mind that health care includes not merely treatments that are curative but also treatments that merely maintain the current health status of the affected person, that slow down the progressive deterioration of the individual's health status, or that are merely palliative in nature. Therefore even if, because of resource constraints, it is impossible for society to fund the effective treatment for certain health conditions because they rank lower in priority than other conditions, the right to treatment—i.e., the right to health care—still remains. It merely means that the right may have to be met in a different way. For instance, it may be met by providing palliative care or by engaging in research—as will discussed more fully in Chap. 5.

[23] Cf. Chap. 2, p. 33 ff. *et pass.*

3.4 Implementing the Right to Health Care

With this as background, what follows is an attempt to sketch the logical structure of a framework of a system of health care that is grounded in the thesis that health care is a right and that tries to incorporate the various ethically based considerations that have been outlined so far.

3.4.1 Some Preliminary Observations

However, to set the parameters for the sketch, it may be useful to re-emphasize certain points. The first is that no matter what view of the right to health care is adopted and irrespective of whether society is clustering or fusional in nature, health care services are commodities. The second is that on a fusional view of society that understands health care to be a right, these services should be defrayed by society as a whole and not on an individual and case-by-case basis by those who access the services. Therefore, ultimately the health care system that is instituted by a society that adopts a rights-model will be a publicly funded system.[24]

A further important point is that as a matter of general principle, the purpose of a publicly funded health care system is not to generate revenue for the society but to meet the health-needs of its members. This does not mean that society may not purchase the services that it is duty-bound to provide from an agency or from individuals who would provide these services as a source of revenue for themselves. All it means is that society itself may not approach it from a revenue-making perspective.

At his juncture, it is important keep in mind that, as was emphasised before, even a publicly funded system cannot ignore outcome and cost/effectiveness considerations since these ultimately influence the resources that will be available to the society and therefore will determine the amount and types of services that it can provide, as well as their distribution. Moreover, while there are certain common factors relative to health care needs that apply to all members of a society, not every member has exactly the same needs. This means that if a health care system is to stand the test of equity, such differences will also have be acknowledge when it comes to health care decision-making, and that they will have to be acknowledged both at the level of policy as well as at the level of individual care.

This, in turn, introduces two further considerations: one that centres in the relationship between policy and hands-on implementation, and one that centres in the need for data in decision making. Both observations are obvious once they have been stated, but they are all too easily forgotten. As to the first, it is essential that the ethics of policy making be consistent with the ethics of the hands-on implementation of those policies. In other words, it would be ethically inappropriate to have a health care system where there exist what may be called a vertical conflict between the ethics of policy and the ethics of hands-on implementation. If there were such a conflict,

[24] For more on this, see Chap. 4, *infra*.

the health care system itself would be ethically inconsistent. In more specific terms, this means that it would be unacceptable to adhere to utilitarian considerations at the level of policy while adhering to deontological principles at the level of hands-on health care practice.

As to the second consideration, rational decisions at the policy level cannot be made without appropriate data. The same thing also applies to rational decision-making at the hands-on level. Data—and this means trustworthy data—are needed at both levels. However, the data that are relevant at the two levels are not exactly the same. At the policy level, data that identify types of issues are important, whereas at the hands-on level data that identify specific issues are important.[25]

However, this does not mean that the data that are needed at the hands-on level of actual practice are limited to the data that derive from the individual patients who present for health care services. To assume this is to ignore the fact that an individual patient's symptoms and conditions are meaningful not merely in reference to the condition of the individual patient but acquire part of their significance in reference to similar symptoms and conditions that are presented by other patients. It is also to ignore the fact that the treatment modalities that are offered for whatever condition with which an individual patient presents are developed through experimentation with groups of individuals.

This essentially means—there are some exceptions, such as the human body's reaction to specific chemicals such as potassium cyanide, which are grounded in the laws of nature—that the likelihood of the effectiveness of a given intervention is determinable only on a statistical basis. Therefore both specific and statistical data are relevant at the various levels that are involved when crafting a health care system that is defensible and workable within the framework of a fusional view of society: in other words, of a society that considers health care to be a right and that therefore structures its operational framework in keeping with the deontological ethical principles that have been identified in the preceding discussion. It is just that the nature of these data is not quite the same.

Also, it is worth re-emphasising the role that the Principle of Autonomy plays in health care. More specifically, while a fusional view of society has a duty to provide health care and a duty to implement, when appropriate, administrative stipulations that mandate how such care may be accessed by its members, such stipulations are not absolute. In other words, this does not necessarily mean that the members of that kind of society are duty-bound to access these services: not even when their conditions are likely to interfere with the well-being of others and the Principle of Nonmaleficence would be engaged. By the Principle of Autonomy, they may choose to absent themselves from interacting with other persons, and the Principle of Nonmaleficence would thereby be satisfied. All other things being equal and with very few exceptions—the exceptions will be explored further in the next chapters—the duty lies entirely on the side of society, not on the side of its members.

Two final points that deserve mention before proceeding any further: *First*, that providing health care services is not simply a matter of providing material items;

[25] For more on this, see *infra*, Chap. 5, "Privacy in Mediated Health Care.".

and *second,* that the right to health care is effectively meaningless unless the relevant material items are made available to those who have the right to health care.

That is to say, providing health care is not simply a matter of providing access to institutions, equipment and the like. While these are necessary components of the matrix that is a health care system, they would be useless and providing them would be pointless if there were no professionals who prescribed them, used them and kept them functioning and operational.[26] And the issue here is not simply one of appropriate funding of the work in which these professionals engage. It also extends to their education and training. Moreover, even with respect to material items, the matter is not confined to providing the necessary (and appropriate) items that have direct health care relevance such as diagnostic devices and the like. It also includes electronic communication and data storage equipment without which modern health can longer function.

At the same time, while health care professionals and the material items in the sense that was just mentioned should be considered an integral part of providing health care, simply providing these without making medications and related items that are prescribed—in other words, items that are used by patients in actualising their right to health care—available on an equitable basis would also not fulfil the right to health care. Pharmaceuticals and the like—and with it the industrial base that produces these items—are also implicated when developing a framework that actualises the right to health care. If these are missing, then an integral component of actual care would be missing. Consequently a society that recognizes a right to health care has to expand its actions into this extended domain as well.

With the exception of the last consideration—it will be dealt with more specifically in the Appendix—what follows is an attempt to incorporate these considerations into the framework of a publicly funded health care system.

3.4.2 Structuring the Framework

It should be clear from the outset that attempting to translate the obligation to provide health care services into an actual framework cannot be done without coordinating individual health care needs with a service structure that is designed to satisfy those needs. It should also be clear that it would be impossible for a society to achieve this on a case-by-case basis and without anticipatory planning, because the ability to meet individual health care needs assumes the existence of an overall system within which the relevant services can be provided. This in turn means that the first step in meeting its obligation is for society to develop an appropriate and functioning policy framework that identifies the typical types of health care needs of its members.

[26] This may become different when artificial intelligence in the full-blooded sense of the term and what might be called robotic instrumentation become a reality. In that case, the robotic entity would replace the health care professional. However, even then, the logic of this consideration would apply in their case.

3.4.3 Duty and the Role of Effectiveness

At the same time, an ethically defensible health care framework cannot ignore reality—and that includes the availability of financial resources—because developing something that is theoretically defensible but not workable in practice is not to fulfil an obligation. Therefore identifying patterns of need is only one step. The ability to successfully respond to these needs also has to be considered. This means that the issue of the effectiveness of the services that are intended to meet that need also enters the picture.

To put it differently, society does not meet its obligation to address the health related problems that undermine the ability of its members to take advantage in an equitable fashion of the opportunities that are available in that society if the services that it provides are not effective. Therefore the issue of effectiveness must necessarily enter into its analysis and planning.

At this juncture it becomes relevant to once more address a consideration that so far has played little role in the preceding analysis. It centres in the question of when a duty has been fulfilled. Initially it was said that a duty is fulfilled when the conditions that are identified in the statement of the corresponding right that gives rise to the duty have been satisfied. This suggests that only an effective outcome—more specifically, only an intervention that successfully deals with the health issue in question—will satisfy the relevant duty. This, in turn, suggests that the failure to successfully address the health needs of individuals who suffer from a certain condition constitutes a failure to satisfy the individual's right to health care.

However, that is not quite correct. The effectiveness of a given intervention is functionally related not only to the condition of a given individual but also to the individual's nature.[27] Therefore it may be impossible to achieve the same outcome for different individuals and groups of individuals even when they are provided with the same service.[28] This, in turn, may suggest that, by the Principle of Impossibility, both the duty and the right, would lapse.

That, however, would be to assume that a duty is fulfilled if and only if the action that is performed in trying to fulfil the duty is successful in dealing with the condition that gave rise to the intervention in the first place. However, as was said, that is not quite correct. To see why this is the case, it may be helpful to step back and once again take a closer look at what is involved in fulfilling a duty.

Obviously, if someone has a duty, then they are ethically obliged to act in accordance with that duty. That almost goes without saying. However, as was just said, whether they are successful in doing so is another matter, because their success may be thwarted by circumstances beyond their control. As was indicated in Chap. 2, the Principle of Impossibility would then exonerate the individual from not having fulfilled their duty.

However, this way of putting the matter is too general because it hides an ambiguity. There are at least three reasons why something may be impossible: It may be

---

[27] Drew (2016), Pirmohamed (2001).

[28] Ali and Al Suwaidi (2019), Wilson et al. (2020), Zhao et al. (2018).

logically impossible, it may be impossible because of the laws of nature, or it may be impossible because of the circumstances that obtain. An example of a logical impossibility would be for a health care professional to both be available for emergency room services and not be available to provide such services. An example of impossibility in the second sense would be the fact that it would be impossible for a single emergency room physician to satisfy the right to health care of two emergency room patients at the same time if this would require the simultaneous presence of the physician at the side of each patient. That would require that the physician be able to be in two places at once, which is against the laws of nature.[29] An example of an impossibility that is grounded in purely factual matters would be an emergency room physician not being able to replace the blood of an emergency room patient who is exsanguinating because the supply of blood is exhausted. Here the impossibility is grounded in the facts of the case.

Nevertheless—and this takes the matter back to the issue of duties—the fiduciary obligation of the emergency room physician towards their exsanguinating patient (which is merely a more specific version of the duty to provide health care[30]) would entail that the physician attempt to deal with situation in some other fashion, such as using substitutes like hemoglobin-based oxygen carriers or perfluorocarbon-based oxygen carriers, etc. If the physician used such substitutes, the attempt to deal with the patient exsanguinating would count as fulfilling their duty even if that attempt was unsuccessful, because it would involve engaging in an appropriate act but which, unfortunately, would have an unsuccessful outcome. The lack of success, therefore, would not constitute a failure on part of the physician to meet their duty to provide appropriate care.

3.4.4 Developing the Framework

Returning, then, to the actual construction of a health service system, such a system would have to be developed in several steps. A preliminary step before even attempting to construct the health care system would be to determine the size of the budget that would be available for society to spend on health care. Without that information, instituting a health care system would be a hit-or-miss affair, and in fact it would not be possible to deliver health care beyond momentary interventions—and even that would be problematic.

The next step in constructing the system would be to identify the life-threatening health conditions that are likely to be found in the society, to determine the number of such cases, and to determine whether according the current state of health science there are interventions that would effectively deal with such cases. The ethical reason for this would be that being alive is a logical precondition for having any right at

---

[29] Laws of nature are not logically necessary. Cf. *Encyclopedia Britannica*, "Laws of Nature;" https://www.britannica.com/topic/law-of-nature.

[30] Se Chapter 5, *infra*.

all and that, since all persons insofar as they are persons are equal, this should take into account all members of society irrespective of age, ethnicity, sex or preferred gender, etc. Continuing, then, with this step, society would have to develop the cost of funding the treatment modalities that presumably would be life-saving or preserving interventions.

However, the right to health care also includes the right to interventions that merely ameliorate health-based suffering even if these are not successful in saving or preserving life. After all, these are health-based conditions that interfere with the ability to lead to a less-than-equal lifestyle and hence, as will be explained more fully later,[31] satisfy the equity condition that governs the right to health care. Moreover, since individuals who suffer from such conditions would still be alive and therefore have a right to health care, society's analysis of projected health care needs would also have to include the likely incidence and prevalence of conditions where no life-saving or –preserving interventions are available and merely palliative treatment is possible. The anticipated cost for all of this should then be deducted from the overall planned budget that was available for delivering other health care.

The next step would then be to identify the statistically average ability of the members of society for health-based reasons to take advantage of the opportunities that society offers, using an objective instrument like the Instrumental Activities of Daily Living (IADL) scale.[32] It would then determine the incidence and prevalence of the types of conditions that interfere with this and determine whether treatment for such condition was possible and effective. Focusing on statistically most common condition, the next step would be to measure the effectiveness of the treatment for this initial condition itself and to express this in terms of an effectiveness range coefficient. The reason for expressing it in terms of a range is that not everyone reacts to treatments in the same way, which is why this coefficient would not have a single value but would indicate a range.

Society would then take the statistically next most common health condition that was likely to impair the ability to take advantage on an equitable basis of the opportunities that society presents and determine the resources that would be necessary to fund treatment for this type of condition—but only if the treatment's effectiveness range for it fell within the range-value of the effectiveness coefficient of the treatment funded for the first condition. The cost of this would then again be subtracted from the total value of the health care budget as adjusted for the cost of the initial type of treatment. If the treatment for this second statistically common health condition did not fall within that coefficient range, the condition would be set aside and the next most common condition would be funded, again using the same criteria. This process would be applied recursively to statistically less and less common health conditions that satisfied the equity criterion and for which treatment existed, and for which the success rate of treatment fell within the effectiveness range that was initially identified.

---

[31] P. 55, *infra*.

[32] Kalligerou et al. (2019), Kojima (2018).

However, this would only constitute preliminary planning for an equitable health care system because it may not only leave a surplus of funds in the overall health care budget—the initial effectiveness coefficient might be too restrictive to exhaust the budget for the services that had been identified on this basis—it would also leave many conditions without treatment as based on their effectiveness coefficients. Therefore the overall effectiveness coefficient's range would be adjusted and, using this modified range, the same process would be run through for types of conditions that had not been funded when using the original coefficient range. The cost of planning for the health care services under this modified coefficient would then be subtracted from the health care budget that was left over from subtracting the originally identified services. This process of modifying the effectiveness coefficient and applying it to as yet unfunded conditions would be repeated recursively until the budget was exhausted. This would effectively address the health care needs of most members of society.

Once again, however, this would only be part of the preliminary planning because it would only address the health care needs of most of society. It would not deal with the health conditions that satisfied the equity criterion for health care but where the otherwise available treatment modalities did not meet the effectiveness coefficient range that had been arrived at using the process that has just been outlined. Therefore, before even engaging in the calculations that have just been outlined, the initial overall health care budget that was made available for funding health care would have to be divided into three parts: one part dedicated to supplying the services that were just outlined, another part that was dedicated to research and to the development of treatment modalities for conditions that would not be covered by the system that was structured in the way that was just outlined, and another part that was dedicated to providing palliative care for those conditions where available treatment would not meet the effectiveness coefficient arrived at under the first section.

The reason for engaging in such a revision before finalizing the overall system would be the fact that although certain conditions would not be funded under the first part because treatments for the relevant conditions did not fall within the range of the final effectiveness coefficient criterion, the individuals who suffered from such conditions would still have a right to health care. And even if the Principle of Impossibility was engaged and no treatment for a particular condition fell within the effectiveness range that was ultimately identified by the process that has just been outlined—for instance, conditions like Creutzfeldt–Jakob disease,[33] multiple sclerosis[34] and amyotrophic lateral sclerosis[35]—this would be true only on the basis of the current state of knowledge and not as a matter of logical impossibility or even according to the laws of nature. Also, as will be explained more fully later, the right to health care would still include the right of access to palliative care.

[33] Vacca (2016).

[34] Howard et al. (2016).

[35] Salameh et al. (2015).

3.4.5 *Research and Palliative Care*

This, of course needs explicit and separate justification since neither research nor palliative care seem to be entailed by the Principle of Equality, which is the basis for claiming that there is a right to health care: After all, neither research nor palliative care remove barriers that prevent an individual from taking equitable advantage of the opportunities that are to be found in society.

3.4.5.1 Research

Focusing, then, on the duty to engage in research, it is important to realize that the right to health care is not exhausted by the right to treatments that are currently available in the relevant society. Nor is the corresponding duty to provide health care limited to providing treatments that currently exist. It also includes the duty to attempt to provide care where currently no curative or ameliorative care exists or where society currently has no modalities that deal with such conditions. It therefore includes the duty to develop effective treatment modalities. This follows from the logical nature of a duty as such, and the duty to provide health care in particular.

To see why this is the case, it may be useful to once again take a closer look at the logical natures of rights and duties, and at their interrelationship. As was pointed out in Chap. 1, a right is a justified claim whose precise nature is indicated by the parameters that constitute its particular character. Consequently—impossibilities being excepted—the corresponding duty is fulfilled if and only if the particulars that satisfy the parameters of the right are executed. Therefore if the particulars contained in a right are specific then, all other things being equal, the corresponding duty will be fulfilled if an only if these specific particulars are executed. On the other hand, if the parameters that are contained in the clauses of the right are general, then the corresponding duty will be fulfilled if and only if the general parameters are fulfilled.

However, the right to health care as it was outlined in the preceding discussion is not the right to a specific intervention. It is a general right: It is a right to the array of actions that fall under the rubric of health care itself. Therefore, purely logically speaking, this means that, all other things being equal, this general right will not be met if the actions that fall under the rubric of health care and that are possible are not implemented. More specifically still, the right has the logical form of a universal proposition, and such a proposition is true if an only all particular instances that are entailed by it are true.

However, again purely logically speaking and turning to the case of specific health conditions, the general right to health care includes the right to treatment for specific types of health conditions if these are instances of the types of conditions that meet the equity criterion that was mentioned above. Thus, rare conditions like Creutzfeldt-Jacob disease would here also be implicated, as would cryptococcosis, etc. but so would more common diseases like measles, malaria and chickenpox, and conditions like appendicitis and tonsillitis. All of these diseases or conditions satisfy the initial equity condition that engages the right to health care. Therefore if there is a right to health care, then society has a *prima facie* obligation to provide the appropriate care for these conditions if these meet the equity condition. The only time that this would not hold would be if the Principle of Impossibility was engaged.

However, the Principle comes in three versions: logical impossibility, natural-law impossibility and factual impossibility. If it was logically impossible to develop a treatment modality for a given health condition, then the general duty to provide health care would not apply. Arguably, that would be the case if someone's brain was macerated and the constituent cells had been liquefied. Even if new cells were reconstructed from the cellular liquid, the cells would not be the numerically same but at best be qualitatively identical and analogues of the original cells. Therefore the individual who was so to speak resurrected would also only be a qualitatively identical individual but would not be the same person. Likewise—and this is relevant with respect to the duty to fund health care interventions—it is logically impossible to spend more money or to use more resources than exist. Consequently, there cannot be a duty to provide health care if the resources that would be necessary to do so exceed what is available.

That aside, then, impossibility in the case of treatment for health conditions is either a function of the laws of nature, or a function of the factual state of affairs. As to laws-of nature impossibility, this would also obviate the duty to provide health care if the condition fell under this rubric. However, as was pointed out before, it is unclear whether the laws of nature would prevent treatment even for conditions such Creutzfeldt–Jakob disease, multiple sclerosis or ankylosing spondylitis. Unless that natural-law impossibility could be established, they and other conditions like them would fall under the rubric of interventions that are impossible only because of the current state of affairs: i.e., only because currently no such interventions exist. From this it follows that for society not to provide treatment for such conditions would be ethically excusable only because no treatment modalities currently exist by the laws of nature.

However, if society does not undertake research that is laws-of-nature possible, then society will be responsible for there currently being no relevant treatment modalities simply because society did not attempt to provide health care services that in principle would be possible. In other words, the reason why it is not possible to provide health care for such cases would be the lack of relevant action on the part of society.

At this juncture, the qualifying clause that conditions the Principle of Impossibility—"where the fact of impossibility … is not a result of the voluntary actions of the relevant party in question"[36]—comes into play. If the lack of appropriate and effective treatment is the result of society's lack of research then, by not engaging in research, society has itself brought about the current lack of treatment. Therefore the Principle would not excuse society for not providing the relevant health care because the situation would be the result of the voluntary action on part of society.

[36] Chapter 2, pp. 23, ff.

Consequently, the general duty to provide health care includes the duty to engage in appropriate research in those cases where no appropriate heath care exists.

However, there is more to the matter. Going back to the system of health care services that would be provided on the basis of the effectiveness schema that was outlined above, if the treatment modality that was provided was effective in dealing with the issue and resolved it in a manner that satisfied the equity criterion and did not have medically inappropriate or unsatisfactory side-effects, then the duty to provide appropriate health care would of course be satisfied. On the other hand, if the modality did not satisfy the equity criterion in its outcome or it had medically inappropriate side-effects, then the duty to provide appropriate health care would not be completely satisfied even though it may address the condition to some degree. In other words, in such a case the duty to provide health care would be met only partially. In order to fully satisfy it, it would be necessary to develop a treatment modality that did not suffer from such drawbacks. Consequently in such a case society would once again have an obligation to engage in research to develop an appropriate type of treatment that did not have such deficits.

3.4.5.2 Ameliorative and Palliative Care

As to palliative care, it may once more be useful to begin by clarifying the notion itself. In common parlance, the term is sometimes used to refer to care that is intended to alleviate the pain or suffering of patients who suffer from life-threatening conditions, who likely are in the process of dying and for whom no further curative treatment is possible.[37]

However, that is too narrow an interpretation. The word itself originally came from the Latin word *'palliare'* which means to conceal or cloak. Palliation, therefore, is any action that is intended to deal with the pain or suffering experienced by someone who is afflicted by a certain condition but where no attempt is being made to deal with the condition itself. In other words, it includes merely ameliorative care as well. It is in this extended sense that the term was used in the preceding discussion when it was suggested that part of the overall health care budget should be allocated to palliative care for those conditions where available treatment would not meet the effectiveness coefficient that satisfied the calculations involved in identifying actual health care services.

At first glance, the right to palliative care does not seem to be justified by the Principle of Equality that underwrites the general right to health care because access to such care would not eradicate or minimize and health-based differences that would prevent individuals from equitably taking advantage of the opportunities that society affords its members.

However, such an inference would be based on a faulty premise: It would be based on the premise that the opportunities that are offered by society only involve

[37] World Health Organization, Definition of Palliative Care; accessed 10/07/2020 at https://www.who.int/cancer/palliative/definition/en/.

interactions of a competitive, quasi-competitive or productive nature. That, however, is incorrect. As was indicated in the preceding chapters, persons are social animals whose very ability to realize their nature as persons functionally depends on their interaction with other persons. Therefore the array of opportunities that society offers is not limited to competitive or productive enterprises. It also includes the opportunities to realize their potentials as *persons*, and persons are more than purely cognitive and productive beings. They also have sensations, emotions and feelings. It is these latter that are addressed in palliative care, and they are involved both in the case of the individuals who suffer from health care conditions as well as in the case of their significant others whose emotional and other characteristics as persons are affected by the patients' conditions. Sine palliative care is the health care that responds to the needs that arise in this connection; it falls under the general umbrella of services that are entailed by the right to health care; and that is why the apportionment of sufficient funds to provide for palliative care is ethically mandated for societies that accept the Principle of Equality and Justice.[38]

3.4.5.3 The Claim of Age-Based Priority

Before concluding this chapter, it may be appropriate to rule out an issue that has sometimes been advocated as entailing that under certain circumstances there is a reduced right to health care: To be more specific, the issue of age.[39] The argument here comes in two distinct versions: *that* advanced age makes health care interventions less effective, which means that the right to health care is accordingly reduced, and *that* older people have already had a chance to access the available services whereas younger people have not, which means that younger individuals should take priority over those who are senescent.

The first type of argument is based on a thesis that is ethically defensible: the thesis, namely, that if a particular type of health care intervention is not effective, then it is no longer obligatory to provide that intervention. However, that thesis does not entail the conclusion that therefore an individual who is advanced in age has a reduced right to health care. As should be clear from the preceding discussion, the right to health care itself holds irrespective of the effectiveness coefficients of the interventions that are involved in providing the relevant intervention. What changes is the manner in which that the right to health care should be fulfilled. More specifically, it means that if the effectiveness coefficient of the relevant health care intervention is reduced below the ranges that are identified by the recursive process mentioned in the preceding sections, then fulfilling the right to health care must be shifted to engaging in research and providing appropriate palliative services.

This clearly has implications for health care resource allocation, as will be discussed more fully in the next chapter. Thus, if the Hayflick limit does indeed

[38] A similar concision is entailed by the Principles of Beneficence and Nonmaleficence.

[39] Williams (1997), Daniels (2007), Bringedal (1992).

determine reduced effectiveness in an aging population,[40] then the effectiveness of a given intervention is functionally age-related. This, however, does not entail that age by itself should be a determinant when evaluating right to health care for individuals taken as group because the Hayflick limit does not express itself in the same way for all persons, and not even if they are of the same chronological age.[41] Moreover, effectiveness is multifactorial and involves many health related factors that include such diverse parameters as stress,[42] socio-economic status[43] and even sex[44] and racial background.[45] As well, the calculation of effectiveness coefficients (which are ranges and not simply unique numbers) take the multiplicity of the various determinants including age into account, and therefore already address the issue.

Perhaps even more important than this is that the reasoning that underlies the first type of argument also suffers from the fact that it inconsistently applies the thesis which supposedly justifies the inference, and that it effectively begs the question. The thesis, after all, focuses on the alleged fact that the effectiveness of health care interventions is reduced with advancing age. However, even assuming this to be true—and, as was said, the available data show that this very much depends on the individual in question—the Principle of Equality entails that if the effectiveness criterion was to be used at all, then it should be applied consistently and to all persons. If this was done, then the cohort to whom it was applied would be group of individuals for whom treatment did not satisfy the relevant effectiveness coefficients, not only the group of individuals who have reached a certain age. Hence it would include both young and old persons. To conclude that it applies only to persons of advanced age would be to beg the question.

As to the second argument—that persons of advanced age have already had a chance to access the available services whereas younger people have not, and that therefore those of advancing years have a reduced right to the relevant services—it also fails on two grounds. *First*, it assumes that the right to health care comes with a time gradient; *second*, that the shorter the time-span for accessing a health care modality, the greater the right to access it, and that the longer one has had the opportunity to access a health care modality the less the right to do so.

However, the right to health care is grounded in the Principle of Equality. Neither the Principle itself nor any right that is entailed by it contains a time gradient, whether that be with respect to the existence of a right or with respect to its strength. Therefore a time gradient of either sort would have to be established on independent grounds. The reasoning here does not do so: it merely asserts it. Consequently the inference begs the question.

With due alteration of detail, the same thing is true with respect to another assumption that the argument require: the assumption, namely, that the strength of a given

[40] Gill et al. (2018), Ogrodnik et al. (2019).

[41] Sozou and Kirkwood (2001).

[42] Benham and Charak (2019), Levy et al. (2019).

[43] Marmot (2017), Veisani et al. (2017).

[44] Beale et al. (2019), Fischinger et al. (2019).

[45] Özdemir and Dotto (2017), Saksena et al. (1992), Stephens et al. (1998).

right is functionally determined by how often the right is claimed or by how long the opportunity to claim it exists. Both of these claims, however, would also have to be established on independent grounds; nor is it clear how they could be established. Thus, it seems counterintuitive to argue that if someone has had a heart attack and has been successfully treated for it, then if that individual had another heart attack their right to treatment would rank lower than the right of someone who has not yet had a heart attack. Likewise, it would seem counterintuitive to argue that someone who has led a prudent health-oriented lifestyle and has not had to take advantage of their right to health care would have a lower-ranking right to healthcare than someone who has led an imprudent life-style but simply was younger. That would hold true only if it was assumed that the length of time that someone has to take advantage of their right determines the strength of that right. And that, of course, is also to beg the question.

Finally—and as an aside—the reasoning based on age also assumes that the opportunity to access specific health services is a constant. If it did not make this assumption then, all else aside, the conclusion would not follow. However, as the history of medicine clearly shows, the assumption is mistaken.[46] In fact, if one were to represent the development of types of health care modalities graphically, it would have the shape of an exponential curve. More and distinct kinds of modalities are being developed all the time. Therefore even if one were to grant that the right to health care had a time-gradient and that its strength was inversely proportional to the amount of time one had the opportunity to access the relevant health services, this would not capture all people of advanced age nor would it capture them equally. In many cases, the modalities would not have existed when they were younger, and many of the younger generations would be on the same footing as those who are aged.

3.5 Conclusion

This chapter has attempted to show how and why access to appropriate health care services is a matter of right on part of members of society, and why it is a matter of duty on the part of society to provide such care. To recap, the argument has been that human beings are persons in the sense of biological entities who have the capacity for cognitive awareness and volition, and who can actualize their nature as persons only within a social setting. It was then argued that as a constructive person that is understood in a fusional rather than a clustering sense, society is an entity that is subject to ethical considerations that include both rights and duties. Relative to the latter, and by focusing on the fact that an ethical person should adhere to the Principle of Equality in the framework of its actions, it was suggested that as a constructive person an ethical society therefore has a duty to institute a system of health care that would reduce as much as possible any health-based inequities that would prevent its members from being able to take equitable advantage of the opportunities that it

[46] Porter (2011).

affords them. It was argued further that its members' ability to access such care should be irrespective of their financial status. The analysis was then expanded by sketching the overall framework that such a system of health care would have, identified several conditions that would govern its framework, and discussed several considerations to which such a system could not ethically subscribe. It was also suggested that although society needs calculative tools for developing and maintaining such a system, these tools would have to compatible with ethical principles that grounded the obligation to provide such a system in the first place, and that this requirement ruled out the use of some of the more common tools that have currency in the field of health care.

None of the considerations that have been presented in this analysis have dealt with what might be called the players who would actually be involved in the functioning of such system of health care. For instance, a health care system cannot function without individuals who actually supply the relevant health care services nor, in modern times, can it function unless there are institutions in and through which the services are provided and in which those who provide the services on a hands-on basis ply their trade. Nor, for that matter, has much been said about those who access the services: namely, the patients. And finally, the discussion has not dealt with the issues of resource allocation at the hands-on level of actual service delivery—of who shall be served when not all can be served at the same time. All of these have ethical implications, and all of them would have to be resolved in keeping with the fundamental ethical principles that ground the establishment a system of health care in the first instance if the analysis of the right to health care is to be anywhere near complete. The chapters that follow will attempt to address these issues and to flesh out the general theory that formed the substance of this chapter.

References

Aggarwal, N.K., M. Rowe, and M.A. Sernyak. 2010. Is health care a right or a commodity? Implementing mental health reform in a recession. *Psychiatric Services* 61 (11): 1144–1145.

Ali, M.T., and Suwaidi J. Al. 2019. Racial and ethnic differences in cardiovascular disease and outcome in type 1 diabetes patients. *Expert Review of Endocrinology and Metabolism* 14 (4): 225–231.

Arnesen, T.M. 2011. The value of DALY life: Problems with ethics and validity of disability adjusted life years. *British Medical Journal* 319 (7222): 1423–1425.

Baker, D.R. 1993. Prioritization of health services. The Oregon Basic Health Services Act and its implications for dermatologists and patients with dermatologic disease. *Dermatologic Clinics* 11 (2): 241–249.

Beale, A.L., D.M. Kaye, and F.Z. Marques. 2019. The role of the gut microbiome in sex differences in arterial pressure. *Biology of Sex Differences* 10 (1): 22.

Benham, G., and R. Charak. 2019. Stress and sleep remain significant predictors of health after controlling for negative affect. *Stress and Health* 35 (1): 59–68.

Bringedal, B. 1992. *Distributional Principles in Health Care.* University of California: California Digital Library.

Cacioppo, J.T., & Patrick, W. 2008. *Loneliness: Human Nature and the Need for Social Connection.* W. W. Norton & Co.

Daniels, N. 2007. *Just Health: Meeting Health Needs Fairly*, 162. Cambridge University Press.

Drew, L. 2016. Pharmacogenetics: The right drug for you. *Nature* 537: 560–562.
Fischinger, S., C.M. Boudreau, A.L. Butler, H. Streeck, and G. Alter. 2019. Sex differences in vaccine-induced humoral immunity. *Seminars in Immunopathology* 41 (2): 239–249.
Gill, Z., M. Nieuwoudt, and W. Ndifon. 2018. The Hayflick limit and age-related adaptive immune deficiency. *Gerontology* 64 (2): 135–139.
Gold, M.R., D. Stevenson, and D.G. Fryback. 2002. Similarities and differences in summary measures of population health. *Annual Review of Public Health* 23: 115–134.
Howard, J., S. Trevick, and D.S. Younger. 2016. Epidemiology of multiple sclerosis. *Neurologic Clinics* 34 (4): 919–939.
Kalligerou, F., R. Fieo, and G.P. Paraskevas, et al. 2019. Assessing fusional status using the IADL-extended scale: results from the HELIAD study [published online ahead of print, 2019 Sep 10]. *International Psychogeriatrics*, 1–9.
Kojima, G. 2018. Quick and simple FRAIL scale predicts incident activities of daily living (ADL) and Instrumental ADL (IADL) disabilities: A systematic review and meta-analysis. *Journal of the American Medical Directors Association* 19 (12): 1063–1068.
La Puma, J., and E.F. Lawlor. 1990. Quality-adjusted life-years. Ethical implications for physicians and policymakers. *JAMA* 263 (21): 2917–2921.
Levy, S., I. Heruti, and R. Avitsur. 2019. Physical injury, stress, and health: Protective role of dispositional optimism. *Stress and Health* 35 (3): 267–276.
Marmot, M. 2017. Social justice, epidemiology and health inequalities. *European Journal of Epidemiology* 32 (7): 537–546.
Nelson, R.M., and T. Drought. 1992. Justice and the Moral Acceptability of Rationing Medical Care: The Oregon Experiment. *Journal of Medicine and Philosophy* 17 (1): 97–117.
O'Neill, O. 2001. Practical Principles and Practical Judgment. *Hastings Center Report* 31: 3 (1–23).
Ogrodnik, M., H. Salmonowicz, and V.N. Gladyshev. 2019. Integrating cellular senescence with the concept of damage accumulation in aging: Relevance for clearance of senescent cells [published correction appears. *Aging Cell* 18 (2): e12942]. *Aging Cell* 18 (1): e12841.
Petrou, P., and S. Vandoros. 2018. Healthcare reforms in Cyprus 2013–2017: Does the crisis mark the end of the healthcare sector as we know it? *Health Policy* 122 (2): 75–80.
Pirmohamed, M. 2001. Pharmacogenetics and pharmacogenomics. *British Journal of Clinical Pharmacology* 52 (4): 345–347.
Porter R. ed. 2011. *The Cambridge History of Medicine*. Cambridge University Press.
R Cohen-Almagor, ed. 2000. *Medical Ethics at the Dawn of the 21st Century*, 23–31. New York, NY: New York Academy of Sciences.
Saksena, N.K., M.P. Sherman, R. Yanagihara, D.K. Dube, and B.J. Poiesz. 1992. LTR sequence and phylogenetic analyses of a newly discovered variant of HTLV-I isolated from the Hagahai of Papua New Guinea. *Virology* 189 (1): 1–9.
Salameh, J.S., R.H. Brown Jr., and J.D. Berry. 2015. Amyotrophic lateral sclerosis: Review. *Seminars in Neurology* 35 (4): 469–476.
Sozou, P.D., and T.B. Kirkwood. 2001. A stochastic model of cell replicative senescence based on telomere shortening, oxidative stress, and somatic mutations in nuclear and mitochondrial DNA. *Journal of Theoretical Biology* 213 (4): 573–586.
Stephens, J.C., D.E. Reich, D.B. Goldstein, H.D. Shin, M.W. Smith, et al. 1998. Dating the origin of the CCR5-Delta 32 AIDS-resistance allele by the coalescence of haplotypes. *American Journal of Human Genetics* 62: 1507–1515.
Vacca, V.M., Jr. 2016. CJD: Understanding Creutzfeldt-Jakob disease. *Nursing* 46 (3): 36–43.
Veisani, Y., G. Moradi, and A. Delpisheh. 2017. Effects of socio-economic status inequality on health outcomes. *Archives of Iranian Medicine* 20 (5): 329.
Warren, C., A.P. McGrew, and L. van Boven. 2011. Values and preferences: Defining preference construction. *Wiley Interdisciplinary Reviews: Cognitive Science* 2 (2): 193–205.
Weinstein, M.C., G. Torrance, and A.McGuire. 2009. QALYs: The Basics. *Value in Health* 12 (suppl.1): 55–59.

Williams, A. 1997. The Rationing Debate: Rationing Health Care by Age: The case for. *British Medical Journal* 314 (7085): 820–822.

Wilson, J.B., L.R. Jackson 2nd., F.E. Ugowe, et al. 2020. Racial and Ethnic Differences in Treatment and Outcomes of Severe Aortic Stenosis: A Review. *JACC. Cardiovascular Interventions* 13 (2): 149–156.

Zhao, Y., Y. Wang, and S. Ma. 2018. Racial differences in four leukemia subtypes: Comprehensive descriptive epidemiology. *Sci Rep.* 8 (1): 548. Published 2018 Jan 11. https://doi.org/10.1038/s41598-017-19081-4.

Chapter 4
Applying the Framework

Abstract This chapter deals with the challenges that the right to health care presents at the hands-on level of actual practice. Particular attention is paid to the issue of avoiding vertical ethical conflict between policy and hands-on delivery of health care. Macro-allocation is distinguished from micro-allocation, and an algorithmic approach to micro-allocation is outlined. The role of effectiveness is also discussed, as is the issue of the greatest good for the greatest number.

Keywords Micro-allocation · Effectiveness · Merit · Utility · First-come-first serve

4.1 Introduction

The discussion in the preceding chapter focused on developing a framework for health care services that is grounded in the Principle of Equality. That analysis was at the level of policy. However, formulating frameworks and policies is not the same as applying them in actual practice. The particulars of hands-on reality may present challenges which, although they should be dealt with in a way that is consistent with the logic of the policies and frameworks, require measures that would make the policies and frameworks too unwieldy if they were stipulated at that level and, moreover, would reduce the latter's essential generality. Nevertheless, such refinements are invaluable because, when properly executed, they guarantee that actual practice is consistent with the general principles that structure the framework and its determinative policies. What follows should therefore be seen as an attempt to sketch such refinements.

4.2 Effectiveness in a Different Key

A consideration that should be mentioned before sketching such refinements deals with the issue of effectiveness; not, however, in the sense in which the notion was dealt with in the preceding chapter but in a more applied sense. As it has been considered so

E.-H. W. Kluge, *The Right to Health Care: Ethical Considerations*, The International Library of Bioethics 92, https://doi.org/10.1007/978-3-030-93838-3_4

far, the effectiveness of a given health care modality determines whether the modality should be included in the list of services that should be funded by an ethical society. However, there is another aspect of effectiveness that is distinct from this, and it centres in the result of using such a service in actual hands-on practice.

The point is that while a particular modality may satisfy the effectiveness criterion for inclusion in the services that an ethical society is ethically mandated to provide, this is a characteristic of the modality itself and is general. More precisely, the notion of an effectiveness range as it was outlined in the previous chapter is a statistical notion. It says nothing about the use of that modality on a particular occasion. Not only that: the use of a given modality not only results in a change in the specific condition that justified using it but—because organ systems are not isolated entities but are functionally interrelated—it also affects the individual's overall health status. When these considerations are taken into account, this raises the question of whether individuals should always have access to the modalities that are identified at the framework and policy level no matter what their effectiveness may be on a given occasion.

More specifically still, there are situations where a particular condition *prima facie* calls for the use of a specific intervention in order to deal with that condition so as to satisfy the equity consideration that underlies the establishment of a system of health care in the first place, as this was discussed in the preceding chapter. However—and this this the present issue—if that is the underlying principle that justifies the use of an intervention on a given occasion, then the result on the overall health status of the individual on whom it is used should also meet the equity criterion. Therefore there arises the question whether, if accessing the particular modality raises the overall health status of the individual who suffers from the relevant condition above the average health status of members of society who do not suffer from that condition, accessing it or continuing to access it is still defensible in terms of the Principle of Equality that underwrites the establishment of the health service framework in the first instance. This is not a question that was addressed when determining whether a given modality satisfies the equity criterion for inclusion in the system of health care services because that centred in the general effectiveness coefficient of the modality.

To answer this question, three considerations are relevant. The first is whether an ethical society is obliged to structure its service framework consistently with the ethical principles to which it subscribes, the second is whether an ethical society can restrict access to specific health services that it otherwise provides, and the third is whether an ethical society's obligation to provide health care services should involve considerations of the outcome of accessing such services on a given occasion.

The first question, of course, requires no further discussion. As has been argued sufficiently in the preceding chapter, an ethical society is a society that incorporates ethical principles into its very framework. This means that any rules, policies or services that it provides should be in keeping with these principles.

The answer to the second question immediately follows from the answer to the first: Since all of the services that an ethical society provides should be in keeping with these principles, the conditions under which these services may legitimately be accessed should also be in keeping with these principles, since otherwise the logical organization of its services would be incoherent. Therefore an ethical society should in fact place conditions on how the relevant services can be accessed: namely, they may only be accessed if, and as long as, such access is consistent with the Principle of Equality. Among other things—the matter will be explored more fully in a subsequent chapter—this means that an ethical society would disallow the private provision or offering of the same health care services that are provided by that society.

Answering the third question involves the considerations that draw on the answers to the first two questions. If a society's reason for providing health care is to satisfy the Principle of Equality then, logically speaking, it must not confine the rules that govern the access to such services by considering only the outcomes of a given intervention for the particular condition in principle. After all, the ultimate reason for providing the services is not simply to deal with specific health issues on their own and considered in isolation simply because they are health issues—i.e., because they are health disorders—but to deal with them because they interfere with the ability of the individuals who suffers from them to participate equitably in the opportunities that society offer to its members. That, however, cannot be determined by simply and exclusively considering the results of applying the relevant modalities to the particular conditions themselves. It must be related to their effect on the overall health status of the individuals who receive the interventions.

To reiterate, the specific results of engaging in particular interventions are important because this is what initially justifies the purpose of their inclusion in the overall array of health services that society is obligated to provide and for providing the service on a particular occasion. However, since the ultimate reason for providing health services is to allow members of society to participate on an equitable basis in the opportunities that society offers, the effect on the overall health status of the individuals who receive interventions cannot be ignored when looking at the legitimacy of providing a particular health care service on a given occasion. Therefore the outcome that accessing the service has on the health-based ability of the individual to participate in the opportunities that society offers to its members has to be considered as well. This, in turn, means that if providing the service has the effect of advantaging the individual in participating in these opportunities over other members of society because of a resultant overall increase in their health status, then providing access to the service would violate the equity criterion that otherwise would legitimate the access to it. Consequently, it would be ethically appropriate for society to cease providing such individuals with this the service.

Therefore the fact that a given modality satisfies the effectiveness criterion for inclusion in the types of services that society is obligated to provide does not mean that the modality should automatically be provided on a particular occasion when the triggering health condition arises. The overall appropriateness of providing or of continuing to provide the service has to be evaluated on each particular occasion. Clearly, this has implications when resources are limited and the issue becomes one of allocation—of who shall have when not all can have—which will also be discussed in what follows.

However, once again, that is not the end of the matter. There is a fourth and ethically pregnant question that arises at the hands-on or micro-level: Does the right to health care as it has been outlined so far entail that health focused interventions should also be provided in cases where the use of the intervention is not consistent with the underlying reason for saying that society has an obligation to provide health care?

That is to say, there are situations where engaging an intervention satisfies the effectiveness criterion in the strict sense of that term and does not seem to violate the Principle of Equality in the sense that was just outlined, but the effect of its use on the overall health status of the patient in question is neither positive nor even significantly retards its deterioration or in any way improves the patient's ability to participate in an equitable manner in the opportunities that are otherwise available in society. This raises the question whether in such a case the right to health care mandates that the intervention should nevertheless be used when resources are limited.

The reason this question arises is simply that an ethical society's duty to provide health care is limited in the sense that the society only has a duty to deal with the health-based factors that would prevent its members from taking advantage in an equitable manner for the opportunities that are otherwise available. Clearly, the overall health status of an individual person plays a significant role in this regard. Consequently when resources are limited it is important to ask whether it is ethically obligatory to provide an intervention if doing so does not have a positive effect on the overall health status of the individual, does not maintains the individual's current health status or otherwise does not contribute to the individual's ability to participate in an equitable fashion in the opportunities that are provided by society. If that is not the case, then it seems that there is no ethically valid reason for using the limited resources that society has and for providing in the intervention—at least, not so far as society is concerned. In effect, it would be a violation of the Principle of Nonmaleficence, because it would result in depriving other individuals who are not able to receive the health care that they need and to which they have a right.

Clearly, then, the matter of access to health care services and the issue of effectiveness is substantially more complicated than what was presented in the preceding chapter. There, the issue was a matter of framework or policy; the present issue is a matter of protocol: of applying the framework or policy in actual practice. What follows, therefore, is an attempt to sketch an appropriate allocation protocol that takes into account the factors that have just been mentioned.

4.3 The Allocation of Resources

4.3.1 Preliminary Remarks

4.3.1.1 Emotional Parameters

Before sketching such a protocol, however, it may be appropriate to acknowledge that allocation in general—which is to say, giving a particular item to one party while withholding it from another—may be emotionally very difficult, and that it may be especially difficult in health care because it may mean that the individual who is shut out will suffer because their health care need is not met, and even that they may die. A personal sense of guilt may therefore arise when making such a decision because one may feel responsible for the negative outcome that will result for the other person as a result of having made such a decision. It may even go against the social or cultural values to which the decision-maker subscribes, and which are part of their identity as person.

However, emotions and feelings are not necessarily valid indicators of what is ethically appropriate, and even social or cultural values may have questionable ethical validity. A good example of this is the fact that some people feel that persons of a different colour or gender from themselves do not have the same rights simply because they differ in that regard, and that many societies and cultures approved—and some still do—of slavery[1] and child labour.[2]

The fact is that, as was seen in the preceding chapter, allocation decisions always have to be made in the arena of health care because resources are always limited, whether this be at the level of hands-on action or at the level of policy formulation. Not to make them is to ignore the ethics of what is appropriate, and ultimately is to let what are essentially accidental factors carry the day. If this were to happen, this would mean that the right to health care would be acknowledged in theory but not implemented in practice. Ethically speaking, therefore, there is no choice but to allocate the health care resources that are available—and that includes specific interventions. What is important, therefore, is to allocate on the basis of valid ethical considerations, and to ensure that these are consistent with the ethical principles that ground the right to health care in the first place. Emotions should be allowed to carry the day.

4.3.1.2 The Greatest Good for the Greatest Number

Another issue that deserves attention before actually sketching what an ethically defensible allocation protocol or schema might look like is whether—or to what degree—considerations that take into account the impact that providing health services to a given individual will or is likely to have on society as a whole are ethically defensible.

More specifically, it is sometimes argued that while serving the health care needs of a given individual may be appropriate when the individual is considered in isolation, not to take into account the overall effect that this would have on the welfare of society as a whole would be to ignore the fact that all individuals are embedded in a social context, and that expending the resources that are necessary to deal with a particular individual's condition may well have a negative effect on society's ability to provide health care for a greater number of people. Further, that in any case the

[1] International Labor Organization (2017).
[2] UNICEF (2019).

obligation of health care professionals is to "do the greatest good for the greatest number," which is crucial when choosing who should be treated first.[3]

However, w hile such considerations may have tremendous emotional and political appeal, they have probative value only if the underlying frameworks that justify them—namely, utilitarianism and related ethical approaches—are ethically valid. As should be clear from the discussion in Chaps. 2 and 3, not only is that extremely questionable, it also guarantees that there will be an irresolvable logical conflict between policy decisions and the ethics that governs the interactions between individual health care professionals and individual patients. Given the importance of this issue, therefore, it seems appropriate to expand a little further on the subject.

As was pointed out in Chap. 2, rights are justified claims that belong to individuals who fulfil the conditions for having them. Thus, the right to vote belongs only to individuals who have the capacity for cognitive awareness and volition and who satisfy the legal requirements for voting. Likewise, and leaving purely legal considerations aside, the right to be treated on an equitable basis belongs only to individuals who are persons and only as long as they are persons. Dead individuals—and that includes brain-dead individuals who may be biologically alive but satisfy the Harvard Criteria for brain death[4]—have no such right. And relative to the context of the present discussion, the right to health care only belongs to persons who are members of a society and only with respect to the society of which they are members.[5]

Furthermore, it will be recalled that rights and duties are correlatives: Whenever someone has a right, then someone else has a corresponding duty. Since the right to health care only belongs to someone insofar as they are a member of a given society, the duty to provide health care is a duty that only belongs to that society. That duty, in turn, will only be fulfilled if appropriate services are provided to the individuals who have the right to health care. Moreover, providing these services involves material actions. A society, however, while being a constructive person in the sense that was indicated in Chap. 1, is not a material entity and therefore is incapable of itself engaging in material actions. Ultimately, therefore, it has to delegate the execution of its duties to material persons who can perform the relevant material acts. In the case of the right to health care, these individuals ultimately—institutional persons may be interposed, as will be discussed in Chap. 6—are the duly qualified and accredited health care professionals.

Therefore when these health care professional interact with individual patients, they in effect are acting as society's agents and are responding to the patient's claim of the right to health care. However, as should be clear from the discussion in Chap. 3, the right to health care is not grounded in the Principle of Utility and therefore in considerations of the greatest amount of good for the greatest number of people. It is grounded in the Principle of Equality. Therefore, insofar as they are acting as society's agents, duly qualified and accredited health care professionals cannot act

[3] Christian (2019), Gomersall, and Joynt (2011).

[4] Wijdicks (2018).

[5] Clearly, this raises the question how the right to health care should be understood at a global level. Arguably, such a right will exist if and only if there is a global society.

on the basis of considerations that violate this Principle. To put this in terms that will be explained more fully in Chap. 5, this means that the interaction between them and their individual patients has to be on a fiduciary basis: i.e., by focusing on what is in the best interests of their patients insofar as that is compatible with the Principles of Equality, Beneficence and Nonmaleficence. Only in this way will the ethics of health care at the policy or framework level be consistent with the ethics of hands-on health care delivery. To put it differently, only when the actions of society's agents are logically consistent with the logic of society's duty will the actions of society's agents constitute the fulfilment of society's duty.[6]

From this it follows that when health care resources are limited at the hands-on level and not all who have a right to health care can be treated—or even can be treated at the same time—the decision framework that the professional decision-makers use to determine priority among the competing claimants cannot logically take social utility into account. To do so would be to violate the Principle of Equality and hence to contradict the very logic that underlies the right to health care that grounds society's obligation to provide health care in the first place.

This does not mean that the likely outcome of a given intervention may not be an appropriate consideration, or that the relative severity of a given condition should not be considered. But it does mean that these and similar considerations will be relevant only if they are focused on the right of the individual patient who presents for care relative to the rights of other patients, and that considerations that focus on "the greatest good for the greatest number" must drop out of the picture. In other words, it becomes a matter of evaluating competing rights between individual persons.

4.3.2 Some Unacceptable Criteria

With this as background, it should be obvious that hands-on or micro-allocation cannot ethically proceed on the basis of criteria such as age, social merit, first-come/first-served or related measures. These may have their own attractiveness and at first glance they may seem quite appropriate. However, attractiveness is a matter of perception and, as was reiterated a moment ago, the key always is—or, rather, should be—whether they are defensible in terms of the ethical principles that ground the framework of health care itself.

4.3.2.1 Age

So, for instance, age may at first glance seem to be an ethically valid criterion.[7] After all—so it may be argued—those of advanced age have already lived and have had their chance at accessing the health care resources that are available, and to allow

[6] For a somewhat different critique of a utilitarian approach, see Moskop and Iserson (2004).

[7] Clarke (2001), Wiese et al. (2012), Wlliams (1997).

them to compete for the same resources on an equal basis with those who are younger would give them an extra chance. Moreover, it may be argued that the Hayflick limit essentially undermines even the effectiveness criterion itself.

However, the doubtful legitimacy of this sort of reasoning and of the use of age as a criterion was already discussed in the preceding chapter, and needs no further consideration.[8]

4.3.2.2 Merit

As to merit, to use this as a criterion for deciding which person takes priority in obtaining health care—in other words, whose right to health care has the greater strength—would be to make the strength of a particular claim to health care dependent on the value judgment that assigns this relative ranking of rights, because it is only on that basis that merit-rankings can be assigned.

That, however, is problematic to say the least. In the first instance, and as was already pointed out in Chap. 1, while values legitimately play a role in determining whether a given right will be exercised by the right holder, they play no role in determining whether the individual who has an obligation should fulfil that obligation.[9] The issue at stake, therefore, is not whether an individual whose right claim conflicts with that of another person may voluntarily and on the basis of their own values subordinate their right-claim to that of the other person—as, was also pointed out, it lies in the very nature of a right that right-holders always have that option—but whether the individuals who are faced with the problem of fulfilling competing duties may unilaterally and on the basis of their own value-judgement rank the strength of the competing right claims. As should be clear from everything that has been said so far, that is not ethically appropriate because it does not fall within their mandate as duty-holders. Therefore judgements of merit as decided on this basis should play no role in assigning a priority ranking to rights in general and to the right to health care in particular.

Further, as was also indicated in Chap. 1, the fact that a particular value is held does not necessarily mean that the value is ethically defensible. So, for instance, the fact that an adult person of advanced age who is disabled contends that they no longer deserve anything because they no longer are productive members of their social grouping, or the fact that a particular transgender person believes that they have less value than individuals who have not undergone a gender change, does not mean that these are ethically defensible values, and this holds true irrespective of whether these values are held by the right-holders themselves or by those who have to decide whether to acquiesce in a right claim and thus to fulfil their duty. The same thing is true of values that rank members of one sex higher than members of another.[10]

---

[8] Chapter 3, pp. 56 ff.

[9] Chapter 1, p. 2.

[10] Hyun (2002).

Moreover, and all of this aside, using merit to determine which of the individuals who are competing for the same health care is to assume that all individuals have had the same chance to acquire merit—however that notion may be understood. Clearly, that is not the case. Social and financial factors play a crucial role in this regard, and these are not ethically distributed.

Furthermore, and again all this aside, using merit as a criterion is to assume that all the data that are relevant for determining a merit ranking are available at the time the decision has to be made. While this may seem to be the case when the individual in question has been accorded an important social standing by a democratic process—for instance, when the individual has been elected as a member of a legislative body—this may be nothing more than the reflection of a political process, and in any case is not necessarily ethically justified. Democratic processes have been known to put in place ethically indefensible values such as slavery, racial discrimination and religious intolerance, and they have been known to put in place ethically despicable persons. Moreover, it is to assume that the supposed merit of the individuals is actually known to those who have to make a decision, which is not always the case—especially in emergency contexts.

And finally, it assumes that whoever makes the decision is actually qualified to do so, and this is not necessarily the case either. It is usually health care professionals who have to make the decision who will be treated first when this is an issue, and while health care professionals are highly trained individuals, their training lies in the area of health care, not in the evaluation of merit.

4.3.2.3 First Come-First Served

As to the first come-first served criterion, closer analysis shows that it is also ethically questionable. To begin with, it is based on the assumption that everyone has the same opportunity to access the health services in question within the same time span, and once more that is not necessarily the case. So, for instance, someone who lives in a rural community that has no tertiary level treatment facilities and where the nearest accessible facility is located in a city far away will depend on transportation to access that facility. That, however, takes time. Therefore the rural resident will always arrive later at the health care facility than someone who experiences the same condition at the same time but who lives in the city where the service provider is located. The rural resident will therefore never meet the first come-first served criterion as opposed to the urban resident whose access is not thus delayed.

With due alteration of detail, the same considerations apply in the case of urban residents of low socio-economic status as opposed to individuals who are socio-economically comparatively advantaged and who have access to transportation to take them to a hospital or other appropriate health care facility. The persons of lower socio-economic status would always arrive later than their wealthier compatriots even though they might experience the same condition and experience it at the same time. Consequently if the rank or strength of a right to health care was determined on the basis of first come-first served, their right would always be ranked lower simply because they would always arrive later. And that, in turn, would violate the Principle of Equality that supposedly structures t health care system.

4.3.2.4 The Matter of "Resources"

These, then, are some criteria that are ethically unacceptable when deciding who shall be treated first when not all can be treated at the same time. However, before actually sketching ethically acceptable criteria for allocation at the hands-on level and how the allocation of scarce resources at the hands-on level should proceed, and before considering how this may be integrated into the overall framework of health care services as this was outlined in the preceding chapter, it may be useful restate clearly once again what is comprised under the term 'resources' itself.

That is to say, the discussion so far may have given the impression that what is at issue when it comes to resource allocation is the access to and distribution of health care services in a material sense only, and it only involves the material items that are involved in providing these services. However, in real terms, that is far too limited a perspective. It is always important to remember that health care professionals also fall under the rubric of resources, and so do health care institutions. If any of these are lacking, then health care cannot be provided, and if any of these are in short supply then exclusionary—which is to say, discriminatory—decisions also have to be made. Therefore in what follows the term 'resources' will be understood in this comprehensive sense.

4.3.2.5 The Issue of Decision-Making

Another issue that should be clarified before sketching what an ethically defensible micro-allocation decision making protocol might look like is the question of who should make the decision in situations when competing claims for the health care conflict, and how that decision should be made.

To put this into context, it was argued in the preceding chapters that health care is a right, and that therefore an ethical society should supply its members with such services. It was also pointed out in Chaps 1 and 2 that a right is a justified claim that may be exercised at the discretion of the right-holder. However, as should also be clear from the preceding discussions, the right to health care is not a right that comes without conditions. It is legitimately claimed by an individual on a given occasion only if the individual suffers from a condition that satisfies the equity condition and requires health care to alleviate it, and only for the kind of health care service that is appropriate. Therefore the right-claim comes with what might be called a justification structure, which means that the exercise of the right-claim must be justified in the first instance: there must indeed a health condition that justifies engagement with a particular kind of health care modality, and the direction of the claim—i.e., the claim to a particular type of health care service—must also be justified.

It is at this juncture that the fact that the right to health care is a societally grounded right and that therefore providing health care is a societal obligation again becomes relevant. To once again put the point first in general terms, an obligation is a duty that is owed by someone to someone. A societal obligation, therefore, is an obligation that belongs to society considered as a whole, and it is a duty that is owed by a society to its individual members.

However, as was also said before, a society is not an entity in the sense that it can discharge its duties itself because it is not an entity that can interact with other entities in any material sense. Consequently, if a society wants to discharge its duty of providing health care—which involves material interactions—it must use individuals who can function in that role and who, as its agents, can provide that care. These agents, however—and the ethical issues surrounding health care professionals will be considered more closely in the next chapter—must function as its agents not only in the sense of discharging its hands-on duty of actual delivery but also of deciding whether a particular claim is justified. After all, every duty-holder also has the duty to determine whether a particular right-claim is in fact valid, because only if the claim is valid will the corresponding duty enter into force. This, then, entrains two questions: One is the question of whether health care professionals are the appropriate individuals to make such a decision; the other is the question of how the validity of the claim to health care should be established on a given occasion.

As to the first question, it will be addressed more fully in the next two chapters; as to the second question, here it is important to note that the right to health care is importantly different from some other types of rights. Some rights—for instance, the right to shave one's head or to cut one's toenails—do not depend on anyone else to determine whether claiming them is valid. The right-holders may determine them freely and on their own because they are, so to speak, self-regarding rights and the duty—if indeed that is the correct term—to act in response to the right-claim does not involve other persons but involves only the individuals themselves.

The right to health care is importantly different in this respect. It does involve others because responding to such a right-claim involves providing services that are performed by other persons where—for reasons that have already been indicated—these other persons would effectively be acting as society's agents. These agents must not only provide the relevant service but also assess the validity of the right-claim in the first instance simply because not every such right-claim—i.e., not every claim to health care—is valid. Thus, claiming the right to health care because one is merely feeling ill does not automatically engage society's duty to provide health care because feeling ill is not necessarily caused by a condition that requires medical intervention. Therefore the validity of such a claim has to be assessed, and it is only when a health condition that meets the equity condition is involved that society's duty to provide health care is engaged. However, assessing the validity of the claim involves considering the material facts of each case, and this involves health care professionals.

Not only that: It must also be decided what health care service would be appropriate. The claimant's own request as to what nature that service should take is not necessarily appropriate or decisive. That is to say, the right to health care is the right to an array of health care services in response to a health condition that requires treatment. Historically—and this is certainly true in the contemporary setting—most individuals were aware of a spectrum of health services that are available in their society. Therefore on occasion they may request a particular service in order for their complaint to be assuaged.

When health care is not construed as a right, this might be acceptable because then health care services are commodities on a par with other commodities,[11] and receiving health care amounts to no more than purchasing a commodity. Moreover, as long as the strictures that otherwise apply to the purchase of commodities are met, the issue of right does not really enter the picture. By contrast, when health care is considered a right, society must ensure that the health care services that it provides are used appropriately and effectively because society's resources are limited. This means that access to the services must be controlled, and that the wishes of a particular claimant—i.e., of a particular patient—should not necessarily determine whether and what service should be supplied. In other words, when health care is construed as a right, the decision whether to honour the right-claim does not lie in the hands of the patient who asks for the care, nor will the nature of the intervention be determined by the patient. It lies in the hands of the appropriate health care professional who, as was said, then functions as society's agent,

What follows, then, is an outline of how these various considerations are relevant, and what role they play when it comes to ethically acceptable decision making in the face of limited resources at the hands-on or micro-level. And again, it is worth emphasizing that even though someone's right-claim may take priority, have greater weight or rank more highly than another, this does not necessarily mean that the other individual's general right to health care in lapses. It may merely mean that the temporal framework of providing that particular service is reordered on that occasion.

What follows, then, are some initial considerations that should guide an ethical allocation protocol at the hands-on level.

4.4 Ethically Relevant Criteria

4.4.1 Life Saving

Perhaps the most important parameter in this connection is that life-threatening conditions always mandate treatment priority over quality-enhancing conditions. The reason for this, as was pointed out before, is purely logical and lies in the nature of a right.

To again put it first in purely general terms, since a right is a justified claim, it does not exist in a vacuum. It always belongs to someone. Therefore a right can only exist if there is right-holder who has that right. From this it follows that the

[11] For the sake of brevity, this ignores health care services that were provided by religious societies and similar organisations.

existence of a right-holder is a logical precondition of there in fact being a right at all. Consequently, and by that token, the failure to preserve the existence of the right-holder is also the failure to preserve the existence of the right. Therefore if the right in question is valid and its existence has temporally sensitive implications for the existence of the right holder, and if only engaging in the relevant action at that point in time would underwrite the existence of the right-holder, then that right takes priority over other rights that do not fulfil the same condition.

Transposing these considerations into the present context of health care, this means that when two or more claims to health care conflict but one of the relevant interventions is life-saving whereas the others are merely quality-of-life enhancing, then ethically speaking the life-saving intervention claim would take priority. The other interventions could be postponed until no further life-saving intervention needed to be considered or no other priority-ranking condition exists.

4.4.2 *Possibility*

Once the life-saving vs, quality enhancing priority issue has been resolved, the next point to be considered is the issue of possibility: Is it possible to provide an appropriate intervention for the condition in question?

For reasons that are ultimately grounded in the Principles of Impossibility and Nonmaleficence, the claim where the outcome profile of the relevant intervention has the higher degree of probability would have a higher degree of priority over the one with a lower likelihood of successful outcome. This may mean that if there are several patients who are waiting for treatment, this would involve an iterative process of evaluation. At the same time, if the overall health profile of a patient was such that it was clear that the issue with which the patient presents cannot be successfully dealt with at all, then then patient should be removed from the list of competitors for the scare resources and relegated to the domain of palliative care.

Of course there may be several reasons why it may not be possible to successfully deal a particular condition. For instance, it may in principle be impossible to deal with the condition or it may be impossible to do so because the necessary modality is unavailable and society just does not have the resources to provide it. An example the first would be the sort of case where a patient is exsanguinating and has lost so much blood that resuscitating the individual is impossible even if blood were supplied[12]; and a different sort of case would be that of the patient who has had a mountain climbing accident, suffers from extreme hypothermic and was not recovered until days later.[13] In none of these cases would the decision to rank the right to health care of these individuals lower than that of other individuals and not to provide life-saving services be unethical, because the lives in question simply could not be saved in any case.

[12] Huei et al. (2017). Published 2017 May 31; Rich et al. (2018).

[13] Huey and Eguskitza (2001), Moore and Semple (2013).

As to impossibility grounded in the lack of societal resources, some modalities are so resource-intensive that providing them would essentially make it impossible for society to fund any other health care interventions. For instance, this would be the case if a society was so impoverished that it could not afford such cutting-edge interventions as heart transplants or double-lung transplants or medical treatment of spinal muscular atrophy because that would be at the cost of preventing infectious diseases or of dealing with issues like acute appendicitis.[14]

That aside, therefore, the possibility criterion as thus understood provides a further and ethically acceptable criterion for allocating health care when the question of intervention priority arises. Therefore impossibility in this extended sense provides an ethically valid criterion for ranking the priority or strength of competing health care claims. However, as was noted before, this does not mean that the right to health care itself lapses. It merely means that the claim has less strength or lower priority on that occasion, that it should not be satisfied on a priority basis and that it should revert into the iterative process that was mentioned above.

4.4.3 Need Sensitivity

Once it has been established that a particular condition satisfies the possibility criterion, the question becomes whether the situation is need-driven or more or less optional at that point in time.

In a sense, this matter has already been dealt with in terms of the life-saving versus quality-enhancing criterion; however, at this stage in the protocol the issue is not one of life-or-death but whether the intervention that would be required to successfully deal with the condition had to be engaged here-and-now, or whether it could wait. In other words, at this point the issue is whether the need for intervention is so to speak time-driven.

At this juncture, the Principle of Nonmaleficence comes into play. More specifically, if not providing the appropriate intervention at that point in time for a particular condition would result in serious and irremediable harm whereas postponing an intervention for a competing case would not have that result, then the claim for the former has greater strength and has a higher priority than the claim for the latter.

To see why this would be ethically appropriate, it may be useful to spell this out more fully. Sometimes there is exclusionary competition for health care resources, whether that be for access to relevant health care facilities or to health care professionals, and although the claims may be medically appropriate and the relevant care could be provided, it is impossible to satisfy the competing claims at the same time. At first glance, it seems that those whose responsibility it is to provide care are faced with three options: either to provide care for neither individual because both claims have equal strength but they cannot both be satisfied at the same time, to accord a priority ranking to the competing claims by casting lots to see which goes first where

[14] Stringer (2017).

each claimant is given the same chance of coming out the winner, or to consider which need for intervention is time-sensitive.

As to the legitimacy of drawing lots, this will be explored further a little later. At this juncture, suffice it to say that drawing lost would be ethically appropriate if and only if the two claims in fact are equal and that—as will also be discussed more fully in a moment—may not be the case.

As to treating neither, at first glance this may seem to be justified by the Principle of Impossibility because it is simply impossible to treat both individuals simultaneously. However, two points count against such reasoning. In the first instance, the impossibility of satisfying both claims simultaneously does not logically undermine the legitimacy of either claim nor does it mean that treatment as such is impossible. It merely means that the claims cannot be satisfied simultaneously and that a choice has to be made.

In the second place, to appeal to the Principle of Impossibility at this juncture is too precipitous. It is to ignore the fact that there are several ways of resolving such an impasse. One of these—and indeed the one that should be considered at this point—is to rank the priority of treatment on the basis of need—and the criterion of time-sensitivity does just that: The condition that would irreparably deteriorate if it was not dealt with at that point in time would take priority over the condition that could wait without this resulting in an irremediable deterioration in health status.

It may be useful to briefly point out why the Principle of Equality is not violated by adopting this ranking criterion. It is not violated because the right to a health care intervention is equally recognised in both cases. For, if a particular condition, irrespective of its nature, may be addressed at a later point in time, then not dealing with it at that particular instant and merely moving it lower in the ranking of priorities still retains its status as a specific instance of the right to health care. It simply responds to the claim later. Therefore no matter how serious or trivial the condition may be, dealing with it later and without deficit to the individual who has that condition will satisfy the equity criterion that was outlined in Chap. 3. By contrast, if the condition in question would deteriorate irremediably then not providing appropriate treatment at that point in time would be to violate the equity criterion because it would be to treat the right to an equitable health status differently in the respective cases.

It should also be noted that the time-sensitivity criterion that has just been outline is a purely formal criterion that focuses solely on acuteness of need. It does not mean that other factors may not influence and even change the priority ranking that would be assigned on this particular basis. It merely means that all other things being equal—and that is an important consideration—when two competing claims for health care confront each other, the one that meets the time-sensitivity criterion has greater strength and meeting it takes priority.

4.4.4 Standing

In cases where the life-saving, possibility and time sensitivity criteria do not provide a priority-ranking for the competing claims to health care, a criterion that may offer a way to resolve the impasse is centred in the standing of the claimants: Not, however, in the sense of merit or of being a member of a recognized social or political group as opposed to not having that sort of standing, but in the sense of being an active member of a health care profession that is engaged in providing the kind of care that is at issue. Therefore whether a particular individual was homeless, unemployed or otherwise a member of some minority group—indeed, whether the individual was a member of any group at all, where this includes hereditary or political positions—would be irrelevant as a ranking consideration in this sense of the term. On the other hand, if one of two competing individuals was an active member of a health care profession in the sense that was just indicated whereas the other was not, and if the former being provided with the relevant care would restore their ability to function in that capacity, then their right-claim would take priority over the claim of those where this would not be the case.

The basis for this criterion lies in the Principle of Impossibility. As should be clear from what has been said so far, the right to health care is meaningless unless it can be translated into practice. That, however, is impossible unless there are active health care professionals who provide the relevant services. In other words, these services depend for their existence and availability on the activities of health care professionals who provide them. It follows that keeping health care professionals who otherwise are active in providing these types of services in a functional state is a logical precondition for the very existence of a system of health care that provides such services. Consequently the right of health care professionals who otherwise are active as health care professionals in this sense takes priority over the right of individual who are not active in that capacity.

The qualifier 'active' is here crucial. Not all individuals who are trained as health care professionals or who are members of a particular health care professional association actually practice their profession. Some of them are retired and others have adopted a different pursuit—for instance, law or politics. In their case, the priority ranking that was outlined a moment ago would not apply because they would not be active members of the health care system that gives meaning to the right to health care in general and life-saving health care in particular.

4.4.5 Origin: Auto- Versus Hetero-Induced

While employing the preceding criteria may well resolve the conflict between competing claims to health care, there are occasions when this is not the case. So, for instance, two patients may arrive in an emergency facility and both suffer from life-threatening conditions but only one patient can be treated at that point in time, and neither the possibility, the time-sensitivity nor the professional-standing criterion can resolve the issue. However, the condition of one patient is the result of negligence or deliberate endangerment to themselves whereas the condition of the other is not. In this sort of case it becomes ethically appropriate to consider the origin of their respective conditions. More specifically, the strength of the claim of the individual whose condition is the result of personal negligence, of failing to take due care of their health, of irresponsible behaviour or of similar circumstances would has lower priority than the claim the individual whose condition did not originate in that fashion. To put it in a nutshell, hetero-induced conditions would ethically take priority over auto-induced conditions.

The full analysis of why this criterion is ethically appropriate will have to wait until Chap. 8, which will deal more extensively with the ethical implications of the right to health care for members of society. At this point, suffice it to say that like any right, the right to health care comes with certain conditions. More specifically, it comes with what previously was called a justification structure. The aspect that is relevant in the present context is the one that concerns the individuals who present with competing claims to health care. To put it briefly, the right to health care belongs to individuals only on the condition that they try to take appropriate care of their health so that health care services would not be required unnecessarily. More specifically still, it belongs to them only on the condition that they will not engage in activities that unnecessarily put their health at risk.

The underlying reason for this qualifying condition, in turn, lies in a combination of the \Principle of Impossibility, the Principle of Nonmaleficence and the limiting condition that governs the Principle of Autonomy that was pointed out in Chap. 2.[15]

The Principle of Impossibility is engaged simply because societal resources in general and health care resources in particular are limited and therefore not all claims to health care can be met. Consequently, not taking appropriate steps to safeguard one's health would be to ensure that these finite resources would be engaged when it would not be necessary to engage them—which in turn would mean that society would not have enough resources to provide the services that otherwise it could—or indeed was obligated—to provide.

This, in turn, would directly engage the Principle of Nonmaleficence: By society not being able to provide the relevant services for the people who need them, harm would arise for such individuals. Therefore, by not taking appropriate care of their health, and by not minimizing their need for health care services, the individuals in question would be the ultimate cause of harm to those affected persons.

The Principle of Autonomy is engaged because any person's exercise of autonomy is subject to the equal and competing rights of others. Therefore while members of society are at liberty to engage in any activity they please, that liberty ceases as soon as the activity in which they choose to engage interferes with the rights of others. Therefore if someone engages in an activity that predictably will result in the need for health care, or that unnecessarily increases the likelihood that this will be the case, then that would *prima facie* require that societal resources would have to be expended

[15] Cf. Chap. 2, p. 22.

to deal with the situation. If providing such care would make it impossible to deal with a competing health claim, then providing such care would engage the limiting clause of the Principle of Autonomy: namely "subject to the equal and competing rights of others." Consequently, even if access to the relevant health care would otherwise be a matter of right, the strength of that right would be reduced relative to that of the individual who did not engage in such activities and who nevertheless requires health care. That is why auto-induced conditions as the term was defined above fall into this category: Hence the criterion of ranking access to health care services.

While this reasoning seems fairly clear and while the criterion may be ethically acceptable for ranking priority of access, using it is another matter. For instance, there is the problem of how certain one could be that either of the competing conditions is hetero- as opposed to auto-induced when the patients simply present as patients and nothing is known about them beyond their medical conditions.

Moreover, it is not quite clear that health care professionals are really qualified to make the relevant judgements. After all, their training is in health care, not in the causal analysis of certain activities. Also, to use this as a criterion would essentially be to add another limiting condition to the right to health care itself: namely, that it is restricted by the duty to be responsible in how one takes care of oneself with respect to one's health. And finally, it assumes that everyone has the same opportunity to act responsibly relative to one's health, as well as has the same understanding of what this would involve.

While these considerations outline serious difficulties with using this criterion, they are not difficulties that cannot be met in principle. As to the fact that this criterion would effectively add another limiting condition to the right to health care itself, the validity of this consideration was already addressed in the preceding paragraphs. The limitation inherent in the Principle of Autonomy meets this consideration.

As to the issue of identifying hetero- as opposed to auto-induced conditions when all that is known about the individuals who present for treatment is the nature of their condition, to treat this as an in principle irresolvable problem is to ignore that in many cases the origin of the respective conditions can be deduced without requiring any great causal analysis about how the individuals who present for care acquired that condition. After all, in many cases the individuals who present for treatment do not present on their own but are brought in by persons who either witnessed what happened or who are knowledgeable about the origin of their condition. So, for instance, how the condition of a mountain climber who is brought in by a search-and-rescue team came about can easily be identified on the basis of the fact that the climber is brought in by a search-and-rescue team who can identify how the accident happened; and how the condition of a professional firefighter who fell from a high ladder while engaged in attempting to rescue residents in a high-rise apartment fire came about can easily be determined on the basis of what the first responders who bring the individual into the emergency room say. Unless one is prepared to assume that whoever brings the individuals in for emergency treatment does not know what they are doing or that they are lying—which is something that in itself would need justification—these facts would clearly indicate that the climber's condition was auto-induced whereas the firefighter's condition was not. Therefore the mountain climber can easily be judged to be responsible for their condition and thus meet the auto-induced criterion because mountain climbing is the voluntary engagement in a high-risk enterprise,[16] whereas the firefighter is just as easily judged not to be thus responsible. To be sure, being a professional firefighter is also to voluntarily engage in a high-risk enterprise, but it differs from mountain climbing in that it is a socially mandated profession that itself is justified by the Principles of Beneficence and Nonmaleficence. And not only is it a socially mandated profession, acceptance of the dangers that are inherent in acting as a professional firefighter can also be justified in terms of the Principles of Beneficence and Nonmaleficence, whereas this is not the case with enterprises like mountain climbing.

Somewhat more difficult is the sort of situation where one of the patients who compete for the limited resources is brought in in an inebriated state and with what is clearly in acute liver failure due to alcoholism or a similar behaviour pattern. Fuller discussion of this type of case is again postponed to Chap. 8, where the implications that the right to health care has for members of society will be discussed more fully. For now, suffice it to say that genetic predispositions and social profile notwithstanding,[17] conditions such as alcoholism involve a measure of individual irresponsibility since not all individuals with the same applicable genetic or social profile succumb to such conditions. A degree of choice, therefore, separates it from cases where irresponsibility is not a driving factor.

Moreover, in cases such as these the nature of the origin of the condition itself can reliably be determined by the attending health care professionals who are trained in doing so because appropriate treatment involves making the appropriate diagnosis and identification of its cause. Thus, while there are many contributing causes for alcoholism, the fact remains that alcohol is not like a pheromone where individuals have no control over whether they will react to the substance. Nor do genetic characteristic invariably determine that someone will become an alcoholic—a point that is underscored by the fact that not everyone who has the same genetic endowment and finds themselves in a similar antecedent condition becomes an alcoholic. Identical twins where one is an alcoholic and the other is not are a case in point.

By contrast, a patient who is brought in with acute liver failure which is diagnosed as being the result of an unforeseen reaction to prescribed antibiotics,[18] nonsteroidal anti-inflammatory drugs[19] or anticonvulsants[20] does not fit the auto-induce criterion. In this case, and using the vocabulary that was introduced in this sub-heading, the condition is hetero-induced. Therefore, all other things being equal—in other words,

[16] Langseth et al. (2018).

[17] Ferraguti et al. (2015), Gorini et al. (2013).

[18] Björnsson (2017).

[19] Sarges et al. (2016).

[20] "Anticonvulsants. In: LiverTox: Clinical and Research Information on Drug-Induced Liver Injury." Bethesda (MD): National Institute of Diabetes and Digestive and Kidney Diseases; April 18, 2019.

the patients satisfying each of the other preceding criteria—the claim of the patient who meets the hetero-induced criterion has greater strength and takes priority.

Of course, that need not be the end of the matter because the respective diagnoses as thus presented may be too limited. For instance, the reason why the medications on account of which the patient was in acute liver failure may be that the individual was leading an immoderate or irresponsible lifestyle that necessitated the prescriptions in the first place. Or the patient may have overdosed on the relevant medications. Whether that is the case cannot simply be determined on the basis of the information that would be available to the health care professionals at the time the patient presents. In that case the hetero- auto-induced criterion would not provide an answer to the question of whose claim has greater strength and who should be treated first.

Another sort of difficulty is presented by the sort of case where first responders bring two patients into an emergency room where both patients are in life-threatening conditions as a result of a head-on collision, or even as the result of distinct accidents, as this sometimes occurs in highly travelled areas. In these and similar situations determining whose condition is auto- as opposed to hetero-induced would require forensic analysis, and that is something that is simply not possible at that instant; nor would health care professionals be qualified to make such an assessment. Even the testimony of the first responders or of the police who may be involved in these sorts of cases would not be sufficient to establish with any degree of certitude the origin of the respective situations because, as was said, that would require forensic analysis, since first impressions can be mistaken.

Consequently in cases like these and in similar situations where none of the criteria that have been outlined so far would provide an answer as to whose condition was auto- as opposed to hetero-induced, a different criterion has to be used. After all, a decision has to be made; and the decision has to be consistent with the ethical Principles that ground the right to health care in the first instance.

4.4.6 *Lots*[21]

The Principle of Equality itself points to a way for dealing with this sort of situation: namely, to treat all individuals as equals, to assign each individual an equal chance of being treated and to proceed on that basis. In other words, it is to assign treatment priority on the basis of drawing lots.

To put this first in general terms, the proposal of drawing of lots is generally regarded as an ethically fair and equitable allocation of chance among individuals who otherwise have no significant distinguishing characteristics, and as such is even recognized in law.[22]

[21] Toroyan et al. (2000).

[22] United States v Holmes. 26 Fed Cas 360 (No 15383) (Cir Ct E Dist Pa), 1842.

In the domain of health care, it apparently was first suggested as an appropriate measure to determine access to health care—albeit in the context of testing the efficacy of different medical treatments—in the seventeenth century, when Van Helmont is supposed to have proposed that lots should be cast among possible members of experimental groups to determine who would receive the active treatment and who would receive the placebo,[23] and it has continued to be used in that role into the 20th and twenty-first century.[24]

Clearly, however, the experimental context is different from the current context, where the issue is not research equilibrium but therapy. Nevertheless, the logic of the technique itself when an equal opportunity is assigned to all concerned individuals is the same, and it satisfies the Principle of Equality because no advantage is given to one individual over another. Nor does using the technique require special expertise or knowledge on part of those who use it. Therefore there are no logical or ethical objections to using it when faced with having to make a decision when there are no ethically relevant characteristics that allow one to distinguish between patients whose claim to care cannot be distinguished on any other basis.

And that, of course, is the crucial issue: Are there really no distinguishing characteristics so that lots have to be cast? Answering that question depends on the care that is taken in determining the facts of the case. Assuming that appropriate care has been taken, however, and assuming that there are no other options because all of the other options involve making ethically dubious assumptions or involve accepting ethical approaches that are at variance with the Principle of Equality, then casting lots would be the appropriate way to resolve what otherwise would be an unresolvable situation.

This, of course, may suggest that one should re-examine the acceptability of the analysis in the preceding chapter that was offered in support of the claim that there is a right to health care. Even more strongly still, it may suggest that the analysis in Chap. 2 which supposedly showed that a deontological approach to ethics is the only one that is logically defensible is flawed or otherwise unacceptable. However, unless it can be shown that either these analyses or the current analysis is flawed, the above considerations stand.

4.4.7 *One Versus Many*

The discussion so far has focused on the allocation of resources when the decision must be made between two claimants. In real life, however, the decisions that have to be made may not always be that simple. They may involve having to make a choice between one individual and a group of individuals all of whom have the same right to care but where satisfying the claim of the single individual would make it impossible to satisfy the claims of the group of individuals, or where satisfying the right claims of the group of individuals would necessarily be to deny the right of

[23] Van Helmont (1662).

[24] Medical Research Council (1948), Chalmers (2001).

the single individual.[25] For instance, blood supplies being limited, a decision may have to be made between a patient who is in haemorrhagic shock and who requires a large volume of blood to save their life[26] and three other persons who also require blood in order to save their lives, but where the amount of blood that would be used in the case of the haemorrhagic shock victim would not leave enough blood to save the lives of the other three patients.

In this sort of case, the considerations that were adduced in the previous discussion seem to face the decision-makers with an insoluble conundrum. Intuitively, the decision would seem to favour the greater-number option[27]; yet according to the preceding analysis using the greater number as a deciding factor would not be ethically acceptable because all persons involved have the same right. At the same time, and all of this notwithstanding, it somehow seems wrong to condemn the other persons to death simply in order to save the life of one person.

Closer consideration, however, shows that the schema that was proposed above also offers a solution to this kind of problem. To begin with, to deal with the issue on the basis of intuition would be to use the wrong tool: Intuitions are not the results of reasoned understanding but of psychological and sociological factors.

That aside, however, in order to show why it would be appropriate to deal with the conundrum on the basis of reason it may be useful to once again go back to what was said in Chap. 1 about the nature of rights and to expand on that a little. Rights—and the focus here is rights in the ethical sense—are justified claims that are ultimately grounded in the fundamental Principles of Ethics. As such, they may conflict with each other. So, for instance, the right to self-determination and autonomous behaviour may be limited by the right to the integrity of another person, as this is illustrated in the case of infectious diseases and the health and welfare of other persons. To use a colloquial expression, and to again quote the words of Zechariah Chafee, "Your right to swing your arms ends just where the other man's nose begins."[28] Alternatively, the conflict between rights may be resolved on the basis of logic. So, for example, the right to life is logically prior to the right to a certain quality of life because existence is the logical prerequisite of there being a right in the first place. Therefore rights may be ranked in terms of priority and strength.

Applying these considerations to the matter at hand, what is here at issue is the strength of competing rights. Each of the parties involved has the same right: namely, the right to life. However, when it comes to evaluating the relative strength of competing rights, each has to be evaluated on its own basis and on its own terms. To proceed otherwise would be to treat rights as though they were additive entities and their strength was independent of their nature. As was suggested in the above analysis, this is not the case. Therefore the fact that several individuals have the same right changes neither the nature nor the strength of their respective rights.

---

[25] This has sometimes been referred to as the trolley problem. For its classic presentation, see Foot (1967). See also Thomson (1976). For its relevance to health care, see Andrade (2019).

[26] Patil and Shetmahajan (2014).

[27] Cf. Andrade, *op. cit.*

[28] Chafee (1919).

Ethically, then, decisions about the right of a single individual against the same right of the many have to be reached by comparing the conflicting right claims individually with each other, and by considering whether the content of the competing rights includes a priority provision, as was illustrated by the quality-of-life versus life consideration above. If that was not the case then, as was suggested above, each claimant's right would have be assigned a number and lots would have to be drawn as to which right should be honoured. To be sure, the outcome of proceeding in this way may be counterintuitive and tragic, but two things should again be kept in mind. First, as was said before, intuition does not determine what is ethical. It reflects psychological and sociological factors. Second, tragedy would result even on the one-looses-to-the-many approach: Even if the greater number was saved, the fact that the individual would die would also be tragic. It should always be kept in mind that not everything that is tragic is unethical; and sometimes ethics forces psychologically hard choices.[29]

4.5 Conclusion

The discussion in this chapter has outlined what the right to health care would ethically entail when instituted in practice. This is not to say that the preceding analysis has been complete. Its focus has not been what such health care should look like when the matter is not problematic—for instance, when a members of society present as patients and competition is not an issue. Nor has it dealt with issues such as immunization, to name but one particular health care modality, or the availability of health care services in general as this should be provided by a just and equitable society.

In other words, still other important issues remain. The distinctive roles that health care professionals play as the agents who actualize society's duty to provide health care also deserves close attention, as does the role of health care institutions. And the ethics that surrounds the use of electronic records, expert system, electronic data bases or electronic means of communicational etc. has not even been mentioned. These are increasingly important in the contemporary delivery of health care and as underpinning the frameworks that are currently in place to provide the relevant services.

Finally, the preceding discussion has also assumed that those who have a right to health care have the capacity to decide whether and how to exercise that right. However, that is not always the case, and an analysis of the right to health care would be incomplete if it did not also consider what the right to health means in the case of such individuals would look like. Also, while it was pointed out that the right to health care comes with certain conditions on part of the right-holders, what this means was only briefly hinted at and deserves more detailed consideration. Nor was the issue of who exactly counts a member of society really been fleshed out: i.e., whether only citizens or also non-citizen residents and visitors are included. It was

[29] For more on this, see Chap. 8, *infra*.

hinted at in Chap. 3, but deserves closer consideration, and a discussion of the right to health care would be incomplete if it did not address these and related issues. The next chapter will begin to address these issues by considering more closely how an ethical society would structure its health care system, and examine the ethical implications of the different ways of delivering that care.

References

Andrade, G. 2019. Medical ethics and the trolley Problem. *Journal of Medical Ethics and History of Medicine* 12: 3.

Björnsson, E.S. 2017. Drug-induced liver injury due to antibiotics. *Scandinavian Journal of Gastroenterology* 52 (6–7): 617–623.

Chafee, Z. 1919. Freedom of speech in wartime. *Harvard Law Review* 32: 932, 957.

Chalmers, I. 2001. Comparing like with like: Some historical milestones in the evolution of methods to create unbiased comparison groups in therapeutic experiments. *International Journal of Epidemiology* 30 (5): 1156–1164. https://doi.org/10.1093/ije/30.5.1156.

Christian, M.D. 2019. Triage. *Critical Care Clinics* 35 (4): 575–589.

Clarke, C.M. 2001. Rationing scarce life-sustaining resources on the basis of age. *Journal of Advanced Nursing* 35 (5): 799–804. https://doi.org/10.1046/j.1365-2648.2001.01913.x.

Ferraguti, G., E. Pascale, and M. Lucarelli. 2015. Alcohol addiction: A molecular biology perspective. *Current Medicinal Chemistry* 22 (6): 670–684.

Foot, P. 1967. The problem of abortion and the doctrine of the double effect. *Oxford Review* 5: 5–15.

Gomersall, C.D., and G.M. Joynt. 2011. What is the benefit in triage? *Critical Care Medicine* 39 (4): 911–912.

Gorini, G., A.J. Roberts, and R.D. Mayfield. 2013. Neurobiological signatures of alcohol dependence revealed by protein profiling. *PLoS One* 8 (12): e82656.

Huei, T.J., Y. Mohamad, H.T.C. Lip, N. Md Noh, and R. Imran Alwi. 2017. Prognostic predictors of early mortality from exsanguination in adult trauma: a Malaysian trauma center experience. *Trauma Surg Acute Care Open* 2 (1): e000070. Published 2017 May 31.

Huey, R.B., and X. Eguskitza. 2001. Limits to human performance: Elevated risks on high mountains. *Journal of Experimental Biology* 204 (Pt 18): 3115–3119.

Hyun, I. 2002. Waiver of Informed Consent, Cultural Sensitivity, and the Problem of Unjust Families and Traditions. *Hastings Center Report* 32 (5): 14–22.

International Labor Organization. 2017. Global Estimates of Modern Slavery: Forced Labour and Forced Marriage, https://www.ilo.org/global/publications/books/WCMS_575479/lang--en/index.htm.

Langseth, T., and O. Salvesen. 2018. Rock Climbing, Risk, and Recognition. *Frontiers in Psychology*. https://doi.org/10.3389/fpsyg.2018.01793.

Medical Research Council. 1948. Streptomycin treatment of pulmonary tuberculosis: A Medical Research Council investigation. *BMJ* 2: 769–782.

Moore, G.W., and J. Semple. 2013. A tale of two climbers: Hypothermia, death, and survival on Mount Everest. *High Altitude Medicine & Biology* 13 (1): 51–56.

Moskop, J.C., and Iserson, V. 2004. Triage in medicine, Part II: Underlying values and principles. *Annals of Emergency Medicine* 49: 3, 282–287.

Patil, V., and M. Shetmahajan. 2014. Massive transfusion and massive transfusion protocol. *Indian Journal of Anaesthesia* 58 (5): 590–595.

Rich, C., J. Rayner, and N. Raukar. 2018. Nondisplaced pubic ramus fracture associated with exsanguination and death. *American Journal of Emergency Medicine* 36 (2): 342.e1-342.e2.

Sarges, P., J.M. Steinberg, and J.H. Lewis. 2016. Drug-Induced Liver Injury: Highlights from a Review of the 2015 Literature. *Drug Safety* 39 (9): 801–821.

Stringer, M.D. 2017. Acute appendicitis. *Journal of Paediatrics and Child Health* 53 (11): 1071–1076.

Toroyan, T., I. Roberts, and A. Oakley. 2000. Randomisation and resource allocation: A missed opportunity for evaluating health care and social interventions. *Journal of Medical Ethics* 26: 319–322.

Thomson, J.J. 1976. Killing, Letting Die, and the Trolley Problem, 59 The Monist 204-17.

UNICEF. 2019. Child labour, https://data.unicef.org/topic/child-protection/child-labour/.

Van Helmont, J.A. 1662. *Oriatrike, or physick refined. the common errors therein refuted and the whole are reformed and rectified: being a new rise and progress of phylosophy and medicine, for the destruction of diseases and prolongation of life*, 526. London: Lodowick-Loyd.

Wiese, C.H., C.P. Schepp, I. Bergmann, J.M. Hinz, B.M. Graf, and C.L. Lassen. 2012. Altersrationierung : Mittel zur Ressourcenverteilung im Gesundheitswesen. *Der Anaesthesist* 61 (4): 354–362. https://doi.org/10.1007/s00101-012-2009-1.

Wijdicks, E.F.M. 2018. How Harvard defined irreversible coma. *Neurocrit Care* 1: 136–141.

Wlliams, A. 1997. The rationing debate: Rationing health care by age: The case for. *BMJ* 314: 820.

Chapter 5
Providing the Services

Abstract This chapter examines various options of how a society could fund its health care system, and argues that an ethical society would opt for a socialized system where the only determinant for access to the relevant health services is whether, and to what degree, they interfere with an individual's ability to participate for health-based reasons on an equitable basis in the opportunities that the society offers to its members. Various societal implications in this regard such accessibility and the use of modern technology are also discussed, as is the obligation to engage in appropriate research and training.

Keywords Accessibility · Socialised health care · Research · eHealth · Telemedicine

5.1 Introduction

The previous chapters outlined why an ethical society has an obligation to provide appropriate and equitable access to health care for its members, and considered how that that obligation should be discharged in terms of types of health services at the policy level and of access conditions at the hands-on level. The underlying reason for proceeding in this way was that because a society's resources are always limited, weighted decisions always have to be made both with respect to the kinds of health care services that should be provided and how they should be allocated on an individual basis.

A point that was not addressed in this discussion, however, was how an ethical society should structure the system that provides the relevant services: for instance, whether health care should be provided by a public sector alone, by the private sector where the members of society would in some way be reimbursed for the services that they purchase, or by a mix of public and private sectors, etc. Without addressing this matter, however, the discussion would remain seriously incomplete. The purpose of this chapter is to address this question: i.e., to consider the various possible ways in which an ethical society could structure its health care system, and to examine the ethical implications of the various options among which it has to choose.

E.-H. W. Kluge, *The Right to Health Care: Ethical Considerations*, The International Library of Bioethics 92, https://doi.org/10.1007/978-3-030-93838-3_5

The discussion will draw on considerations that derive from the various aspects that are involved in the actual delivery of health care; i.e., on such matters as financing, the distribution of and access to the implicated services, the training and education of an adequate number of professionals and the organizing of an institutional sector to mention just a few issues, all of which would be integral components of a functional framework that meets not only the pragmatic requirements of an effective health care system but that also satisfies the ethical considerations that were raised in the preceding chapters.

Before going any further, however, it is important to once again clarify a bit of terminology. The phrase "health care" and its cognates is broad in its meaning and in application, and ranges from health services that provides relief from health-based suffering to interventions that respond to beauty- or sports-based needs that have little if anything to do with health in the strict sense of the term that has been identified in the preceding chapters but that involve medical techniques.

To put it shortly, in what follows the phrase "health care" will be used to refer to all and only those services that are intended to address health issues that would impair the ability of individuals to take equitable advantage of the opportunities that are otherwise available to members of society. In other words, unless it is explicitly stated otherwise, the phrase should be understood as referring only to those services that are implicated by the equity condition that grounded the analysis in the preceding chapters, where sports are not included.[1]

5.2 Paying for the Services

5.2.1 *Preliminary Considerations*

Focusing, then, on how an ethical society should structure the funding of the system that provides the relevant health care services for its members, as was pointed out in Chap. 3 there are several possible ways in which this could be achieved. One way would be to treat all such services as commercial commodities[2] and to put in place a framework that allows duly qualified institutions and professionals to offer them on a competitive basis and for society to provide each citizen with sufficient funds to purchase these services as and when they need them.

Another option would be for society to make the provision of the services a government monopoly and to directly fund both the institutions in which they are provided and the professionals who provide them, where access to the services would be free of charge for every member of society: In other words, it would be to institute a system of truly socialized health care in the strict sense of that term.

---

[1] Henceforth, this cause will be taken as understood.

[2] In what follows, the term "commercial" will be taken as understood.

Still another option would be to require all members of society to enrol in a health insurance plan on pain of being penalized for not doing so, where such a plan would include all types of health conditions that would meet the equity criterion that was outlined in Chap. 3, the fees for the plan being centrally controlled by society itself, and where those who could not afford the insurance premiums would be subsidised by society.

A further option would be to institute a system of health expense tax credit, where each individual would pay for the health services that they receive from otherwise private providers but that they would be able to deduct this from the taxes that they would otherwise have to pay. Still other options are possible, of course, as is a combination of any of these, but the preceding should give some indication of the variety of options that would be available to a society to discharge its obligation to provide its members with health care

However, whichever option a society adopts—and there are others—there are two conditions that would have to be met in order for the society to satisfy the ethical conditions that were outlined in the preceding chapters: *First*, the logic of the system that the society puts in place has to be consistent with the underlying ethical reasons for providing health care services in the first place. *Second*, providing the relevant health care services should not be a source of revenue for society that it could use in some domain other than health care.

The first condition needs little comment. If the system of health care that a society establishes did not allow each member of society to have equitable access to health care as this was discussed in the preceding chapters, then the system would not satisfy the Principle of Equality. And if that was the case, then the system that supposedly actualized the right to health care would not fulfil society's obligation to provide health care on an equitable basis.

The second condition deals with the possibility that society might somehow be able to control and limit the costs that would be involved in providing certain types of health care services, and it centres in the question whether it could then treat this surplus amount like "free funds" and, rather than reassigning them within health care itself, assign them for use in some other domain. In other words, as was indicated in Chap. 3, it centres in the issue of whether society may treat the provision of health care as a source of revenue.

As was then said, that would be unethical. The reason for making this claim lies in the logic that triggers the selective allocation of types of health care services in the first place as this was outlined in Chap. 3 when the matter was considered in the context of society's overall obligations. As was then pointed out, selective allocation in any domain of services that an ethical society is obligated to provide is necessitated by the fact that society's resources are always limited. Consequently, a society can apportion only a limited amount of resources to each of the different domains that it is obligated to provide. Therefore each of the various kinds of services that society is obligated to provide must be defrayed within the limits that are apportioned to that domain. This applies to health care as much as to every other domain of service. That is why, because the resource pool that would be assigned to health care would be limited, some health care services would be fully funded in the sense that every effective treatment modality would be funded, but others would not be fully funded, and some would not be funded at all.

However, if it was possible for a society to provide some health care services without spending all of the resources that had been assigned to health care in the first place then, without any change in the size of the overall resource pool that was assigned to health care, it would be possible for society to change the internal allocation process within health care itself. In other words, it would mean that since the original limitation calculations would no longer be valid, these could be recalculated and some of the health care modalities that previously could not be funded at all could now be funded, or perhaps some of the health care modalities that previously could not be fully funded could now be funded in full.

In other words, to treat such a surplus as funds that could be allocated to a domain other than health care would be treat the limited resources that were allocated to health care as though in fact they were not limited: more correctly, as though the initial amount that was allocated to the domain of health care was sufficient to address all health care needs in every way, and as though no choices had to be made. Of course if that was the case—i.e., if somehow the initial overall apportionment to the domain of health care was sufficient to meet all health care needs—it would be ethically acceptable for society to treat the so-called surplus funds in this way. However, given current reality, such an initial sufficiency of funds is not the case.

5.2.2 *Some Options*

That being said, it becomes important to address the question of how the society should structure the funding of the health care services themselves.

Clearly, the option of considering health care services as commodities and maintaining that each member of society should defray their health care expenses as and when they access the services would be an ethically unacceptable option and should be rejected: It would be to deny that an ethical society has an ethical obligation to provide health care services for its members on an equitable basis, because not all members of society are in able to defray the cost of the health services that they need, and therefore it would effectively be to deny that health care is a right. Ethically speaking, one does not have to pay for exercising a right. This option, therefore, need not be considered any further.

5.2.3 *Person Focused Options*

5.2.3.1 Personal Account

Another option that was mentioned was to consider health care services as commodities but to provide each member of society with sufficient funds to purchase whatever services they need. This in turn could be accomplished in several ways. Society could do this by opening a health care account for each individual and placing a sufficient amount into that account so that whatever health care expenses the individual incurred would be covered.

Clearly, however, this would be unrealistic for several reasons. For instance, it would not be known ahead of time what services an individual might require. Moreover, and aside from this, unless society also imposed concurrent controls over the cost of the services themselves, it would be impossible to calculate what the size of the amount should be.

5.2.3.2 Tax Credit

A slightly different and more indirect way of providing each person with sufficient funds to purchase relevant services would be for society to provide these funds by instituting a system of medical expense tax credit, where each individual would pay for the health services that they actually used but would be able to deduct that amount from the taxes that they would otherwise have to pay.

However, this would also not be without ethical problems. While it would not necessarily mean that society would have to impose controls over how much the services would cost, it would mean that that the individuals who claimed such a credit would have to pay for the services themselves first and could only subsequently request the relevant tax credit. That, however, would mean that they would have to have the resources to pay for the services in the first place, which in turn would mean that whole segments of society—e.g., those who live below the poverty line, those who do not have the resources to pay for the services up-front, or those who are otherwise disadvantaged—would not be able to access relevant and appropriate health care. Therefore the equity condition that was mentioned above would again not be met.

5.2.3.3 Insurance and Parallel Stream Option

Another option would be for society to permit the private provision of health services and to shift the onus for funding such care to its members by requiring them to purchase, on pain of a financial penalty, sufficient health insurance to cover their health care needs[3]

This, however, would not be without ethical problems either. For one thing, it assumes that most individuals would have sufficient funds to purchase this insurance. However, if the cost of that insurance was not attuned to the income level of the individuals themselves, this would again run into a problem of inequity, because

[3] For instance, the United States of America *Patient Protection and Affordable Care Act*, accessed 31/08/2020 at https://www.govinfo.gov/app/details/PLAW-111publ148/PLAW-111publ148, would amount to this.

impoverished or otherwise disadvantaged persons would not be able to afford the insurance.[4]

For another thing, unless the cost of the insurance itself was controlled by society, it would not allow individuals to purchase the insurance on a regular basis because insurance costs increase with inflation. Given that inflation rates are variable and generally exceed 0%, this would be problematic because insurance companies are corporate entities that have an obligation to their stockholders to advance their financial interests, and they therefore have to increase their insurance fees to keep up with inflation. So again, the ability of members of society to purchase the necessary insurance would be financial status-dependent, which means that the scheme would again violate the equity condition that mandated the provision of health care in the first place.

Furthermore, since insurance policies are legal documents that specify the specific kinds of services that would be covered, this option would have to assume that the need for health services was reasonably known by the individuals who purchased the insurance. However, this is not necessarily the case. On the other hand, if the insurance that was bought was sufficiently broad to cover all relevant services that an individual might need, this would mean that either the cost for the insurance would become prohibitive for a significant segment of society—something that again would violate the aforementioned equity condition—or some services would effectively have to be provided free of charge. And this, of course, would run into the problem of how insurance companies would meet their obligation towards their stockholders.

In order to avoid these issues, an ethical society would have to supplement any insurance programme of the sort that was just mentioned with a society-funded health care programme that would allow individuals who could not afford the insurance to access that health care as they need it and without seeking the relevant services from the private sector. In other words, it would require the inception of a separate stream of health care services that was distinct from the insurance stream, where this second stream was available to the indigent and to individuals in similar situations. While that would be possible, it would however mean that the health care system that was adopted by the society would involve both a private and a public sector—and that, as will be discussed in a moment—would bring an entirely new set of ethical problems with it.

5.2.3.4 Capitation Option

Still another funding option, as a variant of the personal account option that was discussed above, would be for society to adopt what may loosely be referred to as a capitation model. According to this model a given amount specific to health

[4] Apparently this is one of the reasons why a significant number of individuals do not avail themselves of insurance in accordance with the US *Patient Protection and Affordable Care Act*. See Tolbert J, Orgera K, Singer N, and Damico A. "Key Facts about the Uninsured Population", accessed 2020–09-01 at https://www.kff.org/uninsured/issue-brief/key-facts-about-the-uninsured-population/.

care would be set aside for each of society's members, and the health care services would then be provided by professionals or institutions who could draw on the funds that would thus be set aside for each individual, This amount could either be held by a social agency and be disbursed to the relevant health care professionals or institutions when an individual accessed their services, or it could be held by the individuals themselves and could be drawn on to defray their health care expenses.

Neither of these alternatives, however, would be without ethical problems. In general terms, both variants of this capitation scheme would run into the problem that the equity condition for health care, which is the overarching ethical condition that governs an ethical society's actions, would not be met unless the health status of all individuals who would have a right to health care and the type of access that they would require was known ahead of time, and this is not the case. And—which is just as important—it would not be in keeping with the equity condition unless the health care needs of each individual remained within the parameters that existed at the time of the initial capitulation assignment.

In other words, because the amount that would have to be allocated would have to be set aside for each individual on a prospective basis, it could not be responsive to someone's unexpected health care needs. To be sure, society could attempt to address this issue by using appropriate statistical measures that would identify and quantify the health care needs that the individuals would be likely to encounter. However, this would still not be responsive to any unexpected health care needs that a given individual might run into. The only way this could be dealt with would be if a special part of the resource pool that existed for funding the capitation scheme was set side under a contingency header for precisely that purpose. But again, it would be unclear what its size should be, and historical comparisons would only give a retrospective statistical indication.

As to the option of attaching a specific amount to each individual and assigning it is as a lump sum to be administered by the professionals when an individual became their patient, quite aside from the issues that were identified in the preceding paragraphs, adopting it may well encounter a problem of human nature. Unless the types of services that would be appropriate for a given individual were clearly stipulated when the capitation amount for that individual was assigned, there is evidence that professionals would be tempted to provide uncalled-for services in order to exhaust the capitation fund that would be attached to a particular patient.[5]

5.2.4 Two-Stream Health Care

It may seem that a way for society to avoid any of these issues would be to adopt a variant of the option that was identified under the insurance option and that was briefly discussed in Chap. 3, and to directly choose a two-stream approach to health care; i.e., to structure the delivery of health care in a way that involves both private

[5] Embrett and Randall (2018), Fuchs (2009).

and social providers as distinct and complementary strands that would be available on the basis of choice and ability to pay. Those who lacked the ability to pay would access the social system, whereas those who could pay for the services could access the private stream if they so wished.

Ethically, however, this would also be problematic. Showing why this would be the case in some ways duplicates the analysis that was presented in the preceding chapter when it was argued that an ethical society has an obligation to provide health care on an equitable basis to its members. However, the discussion then did not actually address the possibility that a society might structure the delivery of health care into both a private and a public sector as distinct and complementary systems. Since this approach is globally somewhat prevalent, it seems appropriate to give it special attention. Before doing so, however, it may be useful to outline clearly what a two-stream approach to health care would actually involve.

More specifically, then, a two-stream approach to health care would be an approach where health care services would be available from both a private for-profit sector and from a public or what was called a socialized sector. On this model, the former would provide care on a fee-for-service basis and therefore would provide such care to individuals who could afford their fees, whereas the publicly sector would provide health care services to individuals who either could not afford the fees that would be charged by the private sector or who did not wish to avail themselves of private service option. In each case, however, and with some exceptions, the types of services that would be provided would essentially be the same.

5.2.4.1 Arguments Pro

Some of the more common arguments that have been advanced for instituting such a two-stream approach are that having a private stream would reduce the demand on the public system because it would divert part of the existing demand for health care from the public to the private sector. Not only that: the cost of health care for the public sector would be reduced by such a diversion simply because the cost of providing health care for those who would be served by the private sector would no longer have to be borne by the public sector. As well, it is argued that the wait-times for accessing health care services would be reduced in the public sector because the number of patients who required care would be diminished because of the diversion of some of the demand to the private sector. Reducing wait-times would also improve the qualitative outcomes of health care interventions, which clearly would be in keeping with the Principles of Beneficence and Nonmaleficence. Moreover, health care professionals—and specifically physicians—who are ethically obliged to always act in the best interest of their patients would be able to do so to a much higher degree because sometimes acting in the best interest of a patient means that a patient's needs should be addressed using the best possible interventions, which is not always possible in the limited-resource public sector.

Moreover—and this brings into focus the question whether the health care professionals who chose to participate in the private sector stream would be unethical in doing so—the fact is that health care professionals are also persons. This means that the Principle of Autonomy applies to them as much as it does to other persons. The Principle of Autonomy, however, is to the effect that everyone has the right to self-determination.[6] Arguably, therefore, health care professionals have the right to decide whether they wish to practice in the public or the private sector and, indeed, whether they wish to apportion their time by working part-time in each sector.

And even if it were argued that health care professionals owe a debt to society for the knowledge that they have acquired in their education and for the practical expertise that they have acquired during their residency by interacting with patients, they could pay off such a debt—if indeed there was such a debt— by partitioning their working time between providing health care on a part-time basis in both the public and the private sector. Or—again assuming that there was such a debt—they could pay it off by working for a certain period of time exclusively in the public sector and then leaving the public sector and going into private practice. Therefore, it would seem that in principle there would be nothing ethically objectionable about health care professionals participating in a two-stream model of health care, and for society to adopt such a model. In fact, it would seem that for society to arrange the delivery of health care in this manner would be for it to fulfil its obligation to provide equitable access to health care in the best possible way.

5.2.4.2 Arguments Con

However, as should be clear from what was said about this option in Chap. 3, closer analysis of these arguments suggests that they are not entirely telling. For instance, focusing first on the claim that wait-times in the public sector would be reduced by the diversion to the private sector, the available evidence shows that public-sector wait-times in fact became longer in nations such as Britain and New Zealand when they instituted a parallel stream health care system and permitted a private sector for health care delivery than when there was only a public sector.[7] Therefore if these data are not merely indicative of isolated incidents but of a general trend of what would then happen, if an ethical society has an obligation to provide its members with appropriate health care as quickly as possible—a matter that will be addressed more directly in a moment—instituting a two-stream health care approach would not necessarily satisfy that obligation.

Turning, then to the claim that shorter wait times lead to better outcomes, while intuitively this may seem to be correct, the available data do not in fact support such a claim. It seems that with respect to outcomes, rather than wait-times themselves, what has a more important effect on the quality of outcomes of health care interventions is the initial health-status of the individuals who receive the relevant interventions; nor

---

[6] Chapter 2, pp. 22 ff.

[7] Tuohy, Flood and Stabile, "How Does Private Finance Affect Public Health Care Systems? Marshalling the Evidence from OECD Nations"; referenced in the CHRF study and accessed 04/09/2020 at http://homes.chass.utoronto.ca/~mstabile/oecd.pdf.

do the available data show that longer wait times inevitably lead to a deterioration of health status of the individuals who have to wait.[8]

The reasons for this may be varied. However, the available studies suggest the issue of deterioration that supposedly is associated with longer wait times can be successfully addressed by having appropriate triage protocols that identify when a time-linked deterioration is likely to occur, and by instituting appropriate protocols that make relevant adjustments in priority ranking. In other words, it seems that it can be addressed by appropriate micro-allocation protocols that are structured along the lines that were outlined in Chap. 4.

Moreover, and all of this aside, the available data suggest that on average the outcomes of interventions that are provided by a public system tend to be better than those that are provided by the private sector, the issue of shorter wait-times in the private sector notwithstanding.[9]

Furthermore, the claim that the burden on a public system would be reduced by diverting part of the demand for health care to the private sector ignores the fact that the private sector would have to be staffed by appropriately qualified professionals who could deliver the relevant services. However, since the number of these professionals is limited—a point that will be addressed more explicitly a little later—they would not be available in the public sector. This, in turn—as again seems to be supported by the data that are available from Britain and New Zealand and that were cited above—would mean that the wait-times would be increased for patients in the public sector. Therefore, again, if extended wait-times do affect the likely outcome of health care interventions as this is claimed[10]—i.e., the longer the wait, the less favourable the outcome—this would mean that the wealthy who could afford to turn to the private sector would have their need for health care better addressed than those who could not afford to pay for the relevant services. This once more suggests that instituting a two-stream system would violate the Principle of Equality which underlies society's obligation to provide health care in the first place. Consequently it is highly doubtful that an ethical society could ethically afford to adopt such an approach.

A further problem that adopting a two-stream approach would raise— and this moves the issue to an entirely different level—is that unless the private sector was restricted to providing only the limited types of health care services that were developed in the private sector alone, instituting a parallel private sector would legitimate the private use of social goods.

That is to say, a privatized health care sector that provided a similar array of services as the public sector could only function to any reasonable extent if it could engage in such things as transfusions, since these are required in major surgery; and if it had access to human organs, since these are required to perform transplants for otherwise untreatable organ failures. However, in most jurisdictions human organs are donated, rather than sold. In fact, the trade in human organs is illegal in most

---

[8] Aakvika et al. (2015), Sari et al. (2007), Manji et al. (2013), Newton et al. (2017).

[9] Lasser et al. (2006).

[10] Yip et al. (2019), Nikolova et al. (2015), Reichert and Jacobs (2018).

countries. In general, then, human organs are essentially donated for use in the public sector. As to human blood, even in jurisdictions where blood may be bought and sold, the majority of blood is donated, and it is not donated on the understanding that the private health care sector might earn a profit by being able to engage in the interventions that it offers for sale and that require blood In other words, most of it is donated on the understanding that the blood will be used to meet the need for it in the health care that is provided by the public sector.

Therefore both organs and blood essentially are public goods, and to institute a private health care sector that had access to them—which, as was said, would be necessary if the private sector was to deliver the same kinds of services as was delivered by the public sector—would be to legitimate their unethical diversion. Nor—all other provisions aside—would this be changed if the private sector paid the relevant public agencies that collected these donations. That would not change the fact that the reason for the donations in the first place was motivated by the Principle of Beneficence, and that the focus for doing so was to benefit other members of society and was not to advance the financial profiles of private health care institutions.

A closely related element that is ignored by arguments that favour a two-stream approach to health care is that a private health care sector has to use the knowledge and the techniques which, if they are scientifically valid, have been developed on the basis of appropriately conducted research and experimentation. That research, however, and that experimentation, even when it is conducted by members of the private sector, draws on knowledge and techniques that historically go back centuries if not millennia, and that were developed using human subjects as volunteers. Therefore, once more, the very functioning of a private sector would not be possible if it did not involve using what essentially are public goods.

Of course there also are other reasons for viewing a two-stream approach to providing health care with ethical suspicion. For instance, there are good grounds to believe that introducing a private sector into what previously has been a purely public sector tends to lower the health status of society as a whole. This was demonstrated by the replacement of the Semashko system in the former USSR by the introduction of a privatized health care sector in order to reduce public expenditure.[11] Nor is this phenomenon is unique to the former USSR but holds across nations.[12] By contrast, the reduction of private health care and the introduction of a public stream in the sense that this was discussed above appears to have an overall beneficial effect on the health status of a country's population.[13]

---

[11] Popovich (2011), World Health Organization, "Rocky road from the Semashko to a new health model," accessed 09/10/2020 at https://www.who.int/bulletin/volumes/91/5/13-030513/en/.

[12] van Doorslaer et al. (2006), Duckett (2005).

[13] Baicker et al. (2013), Sommers et al. (2012).

5.2.5 One-Stream Health Care

An obvious alternative, then, to the two-stream approach of funding health care would be to adopt a one-stream approach: more specifically, to make the provision of health care a societal monopoly and to directly fund both the institutions in which such care is provided and the professionals who provide such care, and where access to the relevant services would be free of charge for every member of society as and when it was needed. This would amount to instituting a system of truly socialized medicine in the strict sense of that term.

Arguably, this approach to health care is the ethically most defensible option for a society because it would satisfy the equity condition that was discussed in Chaps. 3 and 4 in all of its aspects. Further, if the data that were previously adduced are correct, it would also result in better access to health care than would be the case by allowing such services to be provided by both a private and a public sector. Moreover, it would be less likely to result in unmet health needs or in the necessity of forgoing medications or services for financial reasons.[14] As an aside, and from a purely pragmatic perspective, it appears that it would also have the advantage of being administratively more efficient than the private health care sector.[15]

However, this option also comes with certain ethically relevant issues. Among other things, it means that the funding for such a system would have to keep pace not only with inflation but also with changes both in the average health status of the population; and as the 1889/1890[16] and the 1918[17] flu pandemics as well as the recent COVID-19[18] pandemic illustrate, this latter may have radical implications for the health services that are required for a given population. Moreover, new tools and techniques for dealing with health issues that require intervention are constantly being developed, and this increases the cost of health care even when nothing else changes.

This suggests that an ethical society would have to consider the allotment of an appropriate resource pool for the provision of health care as a moving target. This would not necessarily mean that the proportion that it accords to health care funding as opposed to the other kinds of services that it is obligated to provide would have to change. However, at a minimum it would mean that the allocation calculations for types of health care at the macro- or policy level would have to be reconsidered on a continual basis if they are to reflect the resource drain that would be accompanied by the sorts of changes that were just mentioned. It almost goes without saying that such re-calculations could not simply follow economic concerns but would have to be guided by the ethical principles that play such a major role in shaping the organizational framework of an ethical society.

[14] Lasser et al. (2006).

[15] Cf. Woolhandler et al. (2004).

[16] Ewing (2017).

[17] Erkoreka (2010).

[18] Ahn et al. (2020).

5.2.5.1 Education and Training

That, however, would not be the end of the matter. The discussion so far may have given the impression that the issue of resource funding for a public or one-stream approach to health care would essentially involve only material factors such as modes of intervention, facilities, equipment, payment for services, etc. However, that would be to adopt too limited a perspective. Another aspect of resource funding on a one-stream approach over and above the funding of these matters—and indeed a crucial aspect—would be that it would be necessary for society to ensure that there was a sufficient number of properly trained health care professionals to meet the health care needs of the population. This would include general practitioners as well as specialists and other professionals who are involved in the delivery of health care: individuals such as nurses, medical technicians, and so on.

Of course this is something that faces all health care systems irrespective of whether they consist of a purely public sector, a private sector or a mix of both. However, it faces a single-stream public approach to health care with particular importance because central to such an approach is the thesis that health care is a right and therefore should not be commodified. This does not mean that the professionals who are engaged in providing that care should not be paid for doing so. Being paid for providing a service is not to commodify the service unless the service itself is something that is available only on a financial and private basis. So, for instance, the fact that police officers who are employed by a given jurisdiction such as a municipality are paid for their services does not mean municipal policing is commodified. Likewise, the fact the government employees are paid does not mean that the institution that is compendiously referred to as the government is commodified. It merely means that the individuals who actual carry out of the tasks that are involved in governing are paid for what they do.

Still, if the education and training that is necessary to turn individuals into health care professionals was not funded by the society but was left to the private sector, this would effectively commodify the knowledge and techniques that become the tools that are used by the health care professionals when they engage in their profession. With this, a quintessential aspect of health care would become commodified because the private institutions would be treating the knowledge and techniques that they teach as though it was intellectual property that they had developed, that they owned and for which they could charge. However, as was pointed out above, that is not the case. As was mentioned before, the information and the techniques that are involved in such education and training essentially are public goods. Not only were they not developed in the private sector, they were developed as based on millennia of social involvement. And even when they were funded and developed exclusively on a private basis—which is generally not the case—they depended on patients who were involved in Phase II and III research protocols in order to achieve scientific validity.[19] It therefore follows that if an ethical society wished to discharge its obligation to

[19] World Health Organization, *Clinical Trials*; accessed 08/10/2020 at https://www.who.int/health-topics/clinical-trials/#tab=tab_1.

provide health care to its members, then it would have to put in place not merely a structure that provided the relevant services but would also have to put in place an educational system that provided that education and training on a non-profit basis. This would not mean that the individuals who receive education and training in the respective specialities in public institutions would not have to pay for their education and training. It means that the fees that they would have to pay would not be for the knowledge and the information that they receive but for the services of the individuals who provide the information and the training.

The issue of education and training extends still further. Different health conditions require different kinds of treatment, and the knowledge and training that is appropriate for one area is not necessarily appropriate for another. Therefore it also means that there would have to be a sufficient number of specialists in the various areas in order to be able to deliver the kind of care that was required. Theoretically, of course, fulfilling this obligation and of ensuring that there are a sufficient number of health care professionals in general could be met by *fiat,* so to speak—i.e., by compelling a sufficient number of individuals to choose general or specialized health care as a career. That, however, would be to violate the Principle of Autonomy. Ethically, therefore, it could only be achieved by offering inducements that would make health care attractive as a profession. How this could be done will not be pursued further at this stage but will be one of the things that will be discussed in the next chapter when dealing more specifically with health care professionals.

5.2.6 Accessibility

Another issue that would be faced by a one-stream and socialized approach to health care would be accessibility. Not, however, in the sense of there being a sufficient number of duly qualified professionals to meet the health care needs of the population but in the sense of ensuring that different access conditions did not undermine the equity condition of the right to health care that lies at the basis of a public one-stream approach.

The point here is that not all persons live in close proximity to each other. This in turn means that they do not all live near to places where the health services that they might need and to which they are entitled are being provided. Consequently their ability to access these health services varies, depending on where they live and also on whether and what kind of access they have to transportation that would allow them to access these facilities. However, once more, the central thesis that underlies a one-stream and socialized approach to health care is that society has an obligation to provide equitable access to the services that would deal with the health-based issues that interfere with an individual's ability to take advantage of the opportunities that are offered by society. Therefore an ethical society would have the duty to ensure that variations in accessibility conditions did not undermine the possibility of equitable access.

Clearly, there are several possible ways of dealing with this issue. The most obvious one would be to establish health care facilities in a geographically dispersed manner such that the same conditions of accessibility would hold for every member of society no matter where they lived. Another way would be to put in place an efficient transportation system that would equalize the amount of time that it would take to access the facilities that provide the relevant services. A third way would be to combine these two approaches and to supplement them with the use of electronic methods of providing health care: for instance with eHealth, telemedicine and electronically augmented diagnostic and treatment tools.

5.2.6.1 Distributed Approach

However—and this again shows how matters of fact can have ethical implications—these options are not ethically equivalent. Adopting the first alternative would essentially mean that society would have to establish a decentralized system of health care where, on the basis of the statistical analysis of the distribution of health conditions, the health care facilities would be distributed throughout the geographic area of the jurisdiction in such a way that access to them would be similar for all members of society.

This approach, however, while it may well deal with the issue of access time—and even this would not be guaranteed—would run into a serious problem of resource allocation both with respect to financing and personnel. As to financing, it would require the establishment, staffing and upkeep of distinct facilities, all of which would constitute a drain on the limited resources that society has available for health care. And while it would probably allow society to respond in an ethically appropriate manner to the health care needs of its members, it would reduce the overall amount of resources that would be available for providing that care itself—which means that the resource pool that was available for macro- and micro-allocation as this was discussed in Chaps. 3 and 4 would be reduced by a comparable amount and society's ability to fund health care services would become even more limited. With this, the Principle of Impossibility would be engaged. With due alteration of detail, the same considerations would apply in the case of the availability of health care professionals and associated personnel. Therefore, ethically speaking, if there was another way of ensuring equitable access, the adoption of such an approach might well be more appropriate.

5.2.6.2 Centralised Approach

The adoption of centralizing health care delivery would not encounter some of these issues. It would also have the advantage of not duplicating both administrative and personnel requirements. However, it would run into the problem of access time for individuals who did not live in the relevant centralised location or locations. And no matter how efficient the transportation system that a society might put in place would be, that in itself would be insufficient to ensure equity of access to health care because the issue of access involves more than merely the location of treatment facilities and of transportation. It also involves diagnosis.

More specifically, individuals need to be diagnosed first before whatever system of transportation that is in place could be validly engaged to bring them to the location where the relevant and appropriate treatment could be provided. After all, as was pointed out in the discussion of the right to health care itself, having the right to health care does not entail that the individuals who have that right also have the right to access the relevant and therapeutic health care facilities at their own discretion as and when they wish. Ethically, the right to health care interventions would be engaged only if a diagnosis by a qualified health care professional indicated that such access was indicated. Therefore the centralised option could not, ethically, be implemented without being supplemented by a distributed diagnostic system. And that, as will be discussed in a moment, also has important ethical implications.

5.2.6.3 Combined Approach: eHealth and Telemedicine

This brings into play the third option: namely, that of combining a distributed approach with a centralized approach and not only providing an adequate transportation system but also using modern technology like electronically augmented health care diagnostic and communication tools, where these latter would be available on a distributed basis, and if necessary engaging in interventional telemedicine.

However, both the distributed and the centralized option tacitly assume that sooner or later patients and health care professionals have to interact in a way that establishes the fiduciary relationship between patients and professionals that characterizes health care. The third option, if it is understood as above, restructures this. It capitalizes on the fact that health care technology has advanced to such a degree that eHealth[20] and telemedicine[21]—i.e., the use of computer-assisted devices to allow health professionals to actively engage in interventions when the professional is in one location and the patient is in a distant setting—have become a viable way of providing health care.[22] This, however, is based on the assumption that the fiduciary relationship that is quintessential to health care delivery can also be established through mediated contact between prospective patients and professionals[23];

In other words, the assumption that mediated contact can establish a fiduciary relationship between patients and professionals of course is central to this approach. If that were not the case, then a society that adopted an approach that involved this methodology in order to satisfy its obligation to provide health services on an equitable basis for its members might well be adopting such an approach because it

---

[20] van den Heuvel et al. (2018).

[21] Bartz (2016), Ekeland et al. (2010).

[22] Choi et al. (2018), Weinstein et al. (2018), Whitten and Mair (2004).

[23] Kluge (2020), Chap. 1.

was practical and economically feasible, but it would be adopting an approach that lacked ethical validity.

5.2.6.4 The Fiduciary Relationship in eHealth and Telemedicine

In fact, of course, a fiduciary relationship can be established under such circumstances, and society's ethical obligation to provide equitable access to health care would be satisfied even when direct and immediate contact between health care professionals and patients does not exist.

To see why that is the case, it may be appropriate to begin with what occurs in the modalities of health care delivery that have been identified as eHealth and telemedicine. The patient, not being in direct contact with the professional, is represented by digitally encoded images and by their electronic health record (EHRs), where the latter includes the data that have otherwise been generated about the patient and that are relevant when assessing their health status. An EHR, therefore, is an electronic version of a patient's medical history and includes such things as diagnoses, prognoses, medications, treatment plans, immunizations, allergies, laboratory and radiology reports, etc.as well as all administrative clinical data that are relevant to that person's care inclusive of demographic data, progress notes, problems etc.[24] In that sense an EHR is, so to speak, a multi-modal informatic image of the patient.

This, in turn, means that logically and ethically speaking EHRs function as patient analogues,[25] and it is this fact that moves them centre-stage relative to the question whether a fiduciary professional-patient relationship can be established in eHealth and telemedicine. More specifically, in both of these cases the inception of the health care professional-patient relationship occurs when the communication between professional and patient is initiated and an analogue of the patient is developed through the development of an EHR or the individual's EHR is accessed by the professional. Once this has occurred, the health care professional has the same obligations towards the patient as would be the case under normal circumstances even though it is mediated through the EHR.

To substantiate this reasoning, it may be appropriate to consider that a fiduciary relationship is not a relationship that holds between health care professionals and any person. It is a relationship that holds between health care professionals and particular patients, where the latter are defined in the previous sense as living members of the species *Homo sapiens* who have the capacity for cognitive awareness and volition. Moreover, the fiduciary relationship is established under normal circumstances not by the fact that the professional materially interacts with and individual who happens to be a patient—more specifically, who happens to be someone who requires health care—but by the fact that the professional encounters the individual in the professional's role as health care professional. Thus, even though a physician

[24] Centers for Medicare & Medicaid Services, Definition; available at https://www.cms.gov/medicare/e-health/ehealthrecords/index.html.

[25] Kluge (2000).

may encounter such an individual at a party, that encounter would not be a professional role-of-physician encounter but a social encounter, which is why no fiduciary relationship would thereby be established.

It being the case, therefore, that the encounter that establishes a fiduciary relationship is context-specific, and that the context is one of delivering health care, the question becomes whether the context has to be one of direct personal interaction. It is at this juncture that the analogue-role of EHRs enters the picture. In eHealth and telemedicine, the encounter between the physician and the individual does not become a fiduciary relationship unless and until that relationship is therapeutic in nature: more specifically, it does not become fiduciary unless and until both the physician and the subject of the EHR agree or it can reasonably be assumed to have agreed—the setting itself may justify such an assumption—that their interaction should be on a physician–patient basis: In other words, that it would be a health care context. That being the case the EHR, functioning as patient analogue, triggers the fiduciary relationship precisely because, even though mediated through the EHR, it is a health care professional-patient interaction.

Logically speaking, this is not unique. It is similar to what occurs in traditional direct-contact health care when diagnostic health services are outsourced by electronic means to a consulting medical specialist such as a specialist who never directly meets the patient but merely provides an analysis of the data that are contained in whatever record is forwarded by the primary treating physician. The fiduciary relationship of the consulting specialist to the patient is, of course, mediated in a temporal sense because the attending physician who forwards the EHR to the consulting physician has prior access to the record and requests the intervention of the consult. That, however, is not ethically relevant with respect to the establishment of a fiduciary relationship between the consult and the patient. What is relevant is that the relationship is established by the fact that the EHR is a patient analogue, and that it is accessed by the consulting professional in the context of their acting in their professional capacity. Therefore even though a direct and personal interaction between the consulting specialist and the individual who is the subject of the EHR never occurs, the relationship between them is fiduciary in nature.

It follows, then, that the use of EHRs in eHealth and telemedicine suffices to establish the role of individuals as patients relative to health care professionals, and that consequently it is appropriate to say that a fiduciary relationship exists between the professionals and the patients even though their relationship is mediated through EHRs.

5.2.6.5 Privacy in Mediated Health Care

One of the issues that arises in mediated health care that is provided by using EHRs and electronic means of interaction and one that finds no real parallels when such services are delivered through direct interaction between patients and professionals is that of privacy. Moreover, it is one where the onus of solving it falls on the society that attempts to fulfil its health care obligation through eHealth and telemedicine, rather than on the health care professionals who actually provide the relevant services.

This should not be misunderstood. It is an ethical given that the right to privacy has been recognized since the beginning of the formalized health care. All other things being equal—some exceptions will be addressed more fully in subsequent chapters—the interaction between patients and health care professionals is cloaked in a mantle of confidentiality.[26] This means that health care professionals have a duty not to divulge any data or information that they may have become aware of relative to the nature and status of the patient. On occasion this even extends so far as to include the identity of the patient, and to ensure that the patient's privacy is respected in all aspects of their interaction.[27] In other words, it has generally been recognized that patients have an almost absolute right of privacy that ethically structures the patient/professional interaction in all of its aspects.

Ensuring that the patient's right to privacy is respected is relatively easily achieved when patients and professionals engage through direct interaction simply by configuring their interaction in a private setting and in such a way that no-one has access to the setting except the individuals who are involved in providing or receiving the relevant care, and by taking appropriate steps to prevent access to the relevant records and by controlling visual, auditory or electronic surveillance. Even when EHRs are involved—as indeed these days is frequently the case because proper treatment generally transcends the temporal limits of a current interaction and draws on patient data that have previously been collected—it is relatively easy to ensure privacy by taking appropriate material steps to safeguard their privacy, if necessary by encryption.

At this point, however, the important point is that in these cases the duty to ensure privacy falls to the health care professionals and no-one else.[28] The situation assumes new dimensions when health care is construed as a right and with a distributed approach to health care and the patient/professional interaction is electronically mediated. It is not simply that EHRs become an essential feature of this type of interaction—although this raises privacy considerations that cannot be met by the same sorts of techniques or devices that can be used in the case of material records—it is also the fact that the transmission of patient data as these are generated at a distance using electronic means involves non-medical professionals. Moreover, the very fact of electronic transmission itself opens up the possibility of privacy intrusion through the unauthorised accessing of the transmission and perceiving (and even copying) the data. And it is because such a method of health care delivery is instituted by society as a way of absolving its obligation to provide health care to its members that the technical system that makes such health care delivery possible becomes society's responsibility. It is this that adds a new and distinctive element.

[26] United Nations, *Universal Declaration of Human Rights*, Article 12; accessed 04/10/2017 at http://www.un.org/en/universal-declaration-human-rights/; World Medical Association, Code of Medical Ethics, accessed 17/09/2020 at https://www.wma.net/policies-post/wma-international-code-of-medical-ethics/.

[27] Deepika et al. (2010), Greenbaum et al. (2011), Huang et al. (2011).

[28] For a modification of this relative to health care institutions, see Chap. 7, *infra*.

At first glance, this matter could be addressed on the same basis as when the patient/professional interaction occurs on a direct basis. In that sort of situation, it is standardly mandated that the technicians who are involved in the generating of patient data—x-ray technicians, laboratory technicians, etc.—are contractually bound to do their best to ensure that patient data will not be available to anyone other than the health care professionals who are involved in the care of the patients and, under certain circumstances, to the patients themselves.[29] Likewise, it is standardly understood that the health care professionals who are involved in the care of the patient, whether that be directly or in a consultative capacity, have a fiduciary duty to safeguard patient privacy and therefore should adopt appropriate measures to achieve.

However, in the electronically mediated system of health care delivery that involves eHealth and telemedicine these measures would be insufficient for the requirement of privacy. While the informatics professionals who would be functionally involved in the design, construction, functioning, maintenance and other aspects of operating the mediated system would essentially be bound by the same general rules that would bind the informatics professionals who are involved in an un-mediated health care setting,[30] the intervening paths which the patient data have to traverse between patients and professionals exposes the data to privacy risks that find no parallel in the unmediated setting. The question therefore arises not only what measures should be involved to ensure privacy protection but also whether the obligation to institute such measures, both in material and financial terms, belongs to the society that institutes the option of eHealth and telemedicine or to the professionals and institutions who actually provide that care and who function, so to speak, as society's agents.

The issue of privacy as it concerns health care professionals and institutions will be addressed more fully in the chapters that follow. For now, and given that the focus of the discussion is the role that society plays in the delivery of health care and the fact that the mediated system of health care would be instituted by society, it follows that ethically the responsibility falls to society. The health care professionals and institutions who are involved are merely the users of the system: To use the terminology that was used before, they are merely society's agents.

[29] Ethically, patients have a right to access all their data. See Regulation (EU) 2016/679 of the European Parliament and of the Council of 27 April 2016 on the protection of natural persons with regard to the processing of personal data and on the free movement of such data, and repealing Directive 95/46/EC (General Data Protection Regulation) (Text with EEA relevance) accessed at http://eur-lex.europa.eu/legal-content/en/TXT/?uri=CELEX%3A32016R0679. For a fuller discussion, see International Medical Informatics Association, Code of Ethics, accessed 18/09/2020 at https://imia-medinfo.org/wp/wp-content/uploads/2019/08/IMIA-Code-of-Ethics-2016.pdf., and Kluge (2020).

[30] Cf. IMIA, *Code of Ethics, loc. cit.*

5.2.7 Planning and Delivering the Services

Another issue that arises for a society that attempts to fulfil its obligation to provide an equitable system of health care for its members but that was considered in only a limited fashion in Chap. 2 is that it requires data in order to be able to plan for and to deliver such services. More specifically, it requires data about the incidence and prevalence of health conditions in order to be able to determine what particular types of services it would be obligated to provide, and to decide how these services should be configured.

However, society's data needs extend far beyond that, and they are not solely health oriented in nature. For instance, the costs that would be associated with providing the relevant services would also have to be identified, where these costs would range from the remuneration that health professionals and associated individuals could expect to receive to such things as the cost of the materials and equipment that would be necessary to deliver the services. Even institutional and transportational costs (insofar as transportation was identified as a necessary component of the distributed model of health care delivery that was discussed above) would have to be factored in, as well as the costs that would be associated with the administrative structure that made health care possible in the first place. Without such data, an overall health care budget could not be developed and a functioning health care system could not be implemented.

However, these are merely preliminary items that by themselves would not allow society to engage in the kind of decision making that is necessary to determine policies at the macro-level. As was indicated in Chap. 3, if society is to discharge its obligation to provide health care in an ethically acceptable fashion it also has to engage in health related research so that it can consistently discharge its obligation to provide health care in an equitable fashion if appropriate health care modalities do not exist in certain kinds of cases.

Ultimately, the data that are necessary in this connection are data that can only be derived from patient records which have to be stored in data banks, and this once more engages the issue of privacy. This time, however, not at the level of patient/professional interaction but at the level of policy making; and while similar ethical considerations arise in both cases, they do not necessarily arise in the same way.

To put this more concretely, the cost of providing health care can only be calculated if data about the number and types of conditions that are to be treated are known. However, because claims made by professionals and institutions are not always trustworthy, adequate data can only be developed by accessing patient records, because these records alone indicate what services are actually being delivered.[31] However, these records will be unavailable unless they are stored in data banks that are accessible for such purposes. Therefore how patient records are stored becomes an issue for a one-stream approach to health care. To use a metaphor, the spectre of Big Data therefore stands in the wings.

[31] Doan (2011), Faunce et al. (2010), Legotlo and Mutezo (2018).

As was said several times, patient records are surrounded by a sphere of privacy which, all other things being equal, can ethically be penetrated only in the therapeutic context of the patient/health care professional relationship. Accountants, however, who have to calculate the actual expenses that are involved in delivering health care do not function in a therapeutic capacity; and neither do the policy makers who ultimately make the decisions about how a health care system would be structured. Nevertheless, these parties have a need to penetrate the sphere of patient privacy that surrounds patient records since otherwise they do not have access to the relevant data and cannot perform their respective roles. The question therefore becomes whether there is an ethical reason that justifies an intrusion on their part into the sphere of patient privacy, and how their obligation to abide by the relevant privacy provisions could ethically be ensured.

Arguably, their intrusion into the privacy sphere would be justified by the Principle of Impossibility. The reasoning here is rather straightforward—at least when health care is construed as a social obligation. A society cannot discharge its obligation unless it has the tools to do so. As was pointed out above, this involves having access to data about health-related usage and to their financial data implications. As was also said, the basis for this can only be derived from patient records. That is why the Principle of Impossibility is engaged. Therefore if health care is construed as a social obligation, then it is legitimate for society to assume that its members, who expect their corresponding right to health care to be satisfied, have tacitly given consent for accessing their records in order to derive the data that are necessary for society to be able to discharge its obligation and develop and maintain a system that provides that care.

However, once again, there is more to the ethical issue of privacy, and this brings into play the situation that was hinted at by the clause "all other things being equal." It involves the Principles of Impossibility, Autonomy and Nonmaleficence. As to the Principle of Autonomy, as was pointed out in Chap. 2, it is to the effect that every person has the right to self-determination subject only to the equal and competing right of other persons.[32] This applies both to what is done directly to the individuals themselves as well as to their records which, as was argued above, are their informatic analogues. Therefore if a patient's health condition imperils the health of other members of society, then for that patient to insist on record privacy would be to interfere with the autonomy of these other members. Therefore the right to privacy here finds one of its limits. Ultimately, therefore, in concert with the Principle of Nonmaleficence, this gives society the right to engage in such things as contact-tracing in the case of transmittable diseases—i.e., the process of identifying persons who may have come into contact with an infected person and the collection of relevant information about these contacts—without this violating the right to privacy.

As to the Principle of Impossibility, it is also implicated because a society will not exist as a functional interrelationship between persons if its members are not alive or are incapable of interacting because of diseases that become rampant and

[32] Chapter 2, p. 22.

the social framework is thereby rendered non-functional due to uncontrolled and negative health conditions among its members. In other words, if a society has an obligation to provide appropriate health care services, then the society must exist as a functional entity. Consequently the limits that condition the right to privacy that were mentioned a moment ago once more slot into place, because then health measures that are designed to ensure the existence and well-being of society's members are necessary, since otherwise the society will disintegrate. Such things as contact-tracing and related measures that were instituted during the pandemics of previous times, therefore, were ethically justified even when this involved a *prima facie* breach of privacy and occurred without the explicit consent of members of society. This would also be the case at the present time with the COVID-19 pandemic.[33]

As an aside, it should perhaps be mentioned that, with due alteration of detail, this reasoning also applies even when health care is construed as commodities. However it does not apply in quite the same way. More specifically, while the existence of society is of course a necessary precondition for private parties to offer health related services, the obligation to participate in contact tracing and related measures applies to them only insofar as it applies to any other member of society, and the right—indeed, the duty—to breach the otherwise applicable privacy rules applies to them only insofar as this holds true for any other members of society who become privy to a threat to the existence and effective functioning of society itself. With due alteration of detail, this reasoning also applies to such things as treason, abuse and related matters. These matters may become a matter of legal decisions and provisions, but that is another matter.

5.2.8 Research

Another matter that is centred in society's obligation to provide equitable access to health care may loosely be referred to under the heading of research. As was suggested above, and as was pointed out in Chap. 3, an ethical society has an obligation to engage in health-related research when no appropriate or effective ways of dealing with health issues exist.

To fully appreciate what this involves, it is important to consider that there are two distinct types of research. There is hands-on research whose purpose is to develop new treatment modalities or to improve on those that exist; and there is second-hand research whose purpose it is to discover relevant correlations between symptoms, treatment and related issues and thus to improve on or to provide new health care services. Both involve access to patient records simply because the relevant data can be found nowhere else. However, strictly speaking, therapeutic intent in the ordinary

---

33 World Health Organization, *WHO Guidance for Surveillance during an Influenza Pandemic (update 20,127)* 2.2 *et pass.* a22/09/2020ccessed at https://www.who.int/influenza/preparedness/pandemic/WHO_Guidance_for_surveillance_during_an_influenza_pandemic_082017.pdf.

understanding of this term obtains in neither of these cases nor, strictly speaking, does a therapeutic relationship in the normal sense of that term obtain.

That is to say—and leaving second-hand research aside for the moment—it should be clear from the discussion in Chap. 3 that not all health care modalities that are available in a society at a given point in time are equally effective nor, indeed, that there even are modalities that are specific to all health conditions and that will ameliorate or otherwise deal with them in an effective manner. As was also pointed out in Chap. 3, this suggests that an ethical society will engage in research in order to deal with these sorts of conditions.[34] This, so it was suggested, ultimately follows from the Principles of Equality and Beneficence respectively.

Hands-on research—which is to say, ethically valid hands-on research—involves several phases. These go from Phase 0 trials which look at whether the pharmacokinetics and pharmacodynamics of a given intervention warrant continued exploration,[35] but more properly range from Phase I trials (in which safety and side effects are evaluated on healthy volunteers who tend to be paid for their participation) to Phase IV trials,[36] with Phases II to III being the relevant intervening phases in this connection. While all these stages involve human beings, it generally is only Phases II to III trials that involve as active participants individuals who suffer from the condition that is being researched; and while in general the hope is that these participants will derive a therapeutic benefit from the interventions, the direct intent of involving them is not to provide therapy but to test the efficacy and appropriateness of the intervention in question.[37] This, in turn, means that the therapeutic intent that underlies the fiduciary relationship in the normal professional-patient setting would not be present and would not justify access to the patients' records. Consequently, the consent process for participation in ethically conducted hands-on research would include specific consent for accessing the participants' records and for using them in a scientifically appropriate manner.

Turning now to second-hand research—and that is really the central issue for a societal-obligation perspective—this of course would not justify access to patient records because the element of patient-professional interaction is not a feature of such research. However, an ethical society could not meet its obligation to provide health care unless both first-hand and second-hand research was conducted. Therefore if they expect their right to health care to be realised in real terms, the right to health care itself entails that patients have an obligation to relax the privacy conditions that surround their records and that society would have a corresponding duty to put in place appropriate and effusive privacy provisions.

---

[34] Chapter 3, pp. 30 ff.

[35] Fromer (2006).

[36] World Health Organization, *Clinical Trials*; accessed 28/09/2020 at https://www.who.int/health-topics/clinical-trials/#tab=tab_1. Strictly speaking, Phase IV trials are more properly referred to as post-marketing surveillance studies because they ae not controlled trials but rather surveillance of the effects of the intervention on uncontrolled populations.

[37] For a different opinion, see Miller and Brody (2003).

This means that certain privacy-centred conditions would apply even in the case of second-hand research. (1) There would have to be a logical connection between the data that are contained in the records and the reason for accessing the records. (2) The reason for accessing the records would have to be scientifically valid in the sense that the research could not reasonably be carried out in any other way. (3) The reason for accessing the records would have to be logically connected with society's obligation to provide health care or the survival of society itself, and that it would not be a matter of purely scientific interest. (4) Whoever accesses the records would have to be a duly qualified individual whose access protocol had been approved by a duly constituted ethics committee. And (5) whoever accesses the records would be obligated not divulge the data or the identity of the individuals whose records they access. The reason for accessing the records, therefore, would have to satisfy the Principles of Impossibility and Equality. Nevertheless, whatever the conditions, society would have an ethical obligation to engage in or to facilitate research into health care, and therefore would have the right to structure its legal privacy provisions in such a way that access to patient records, whether in electronic form or material in nature, be granted to duly qualified researchers engaged in a duly approved research project that was likely to facilitate society to fulfil its obligation to provide health care.

5.2.9 Remuneration

A subject that falls under the rubric of resources but that was addressed only indirectly in the preceding discussion is whether an ethical society is duty-bound to put in place rules and regulations that are designed to ensure that the material resources that it needs in order to be able to provide appropriate and relevant health care for its members will in fact be available in the first place. The issue assumes particular importance in the context of the pricing of diagnostic modalities and of pharmaceuticals because health care at the hands-on level cannot really be delivered without these. However, it also is relevant with respect to the remuneration that health-care and associated professionals receive for their services.

The subject will be addressed more fully in the Chap. 6, where the specific focus is the remuneration that health care professionals are entitled to expect, and in the Appendix, whose focus is pharmaceutical corporations and the ethical implications of their relationship with a society that construes health care as a right. For now, suffice it to say that the prices that they charge for their services or products may make it impossible for society to defray these costs unless society uses the means that are at its disposal to control these prices.

When considering the ethical implications of this state of affairs, it may be relevant to note that as the overarching legal and administrative authority of the social framework, society controls not only which social interactions are permissible but also how they are to take place. Likewise, it also lies within society's legal powers to control the margin of profit that those who offer their products or services may expect from them by making them so-called regulated industries.[38] Therefore if a particular product or service becomes too expensive for society to defray from its limited resources, then it lies within society's legal powers to control the margin of profit that their purveyors may gain.

Furthermore, societies even have the legal right to ignore any patent rights that corporations may have in the products that they sell. After all, it is these patent rights that ultimately prevent any competition between companies, where this might result in lower prices. As will be discussed more fully in the Appendix, the Trade-Related Aspects of Intellectual Property Rights (TRIPS)[39] of the World Tre Organization which is legally binding on all member societies, stipulates that a society may invalidate such rights if it can show that the "legitimate interests of third parties" would be impaired if it did not do so, where the "legitimate interests of third parties" is not to be construed in the sense of what laws exist in the relevant jurisdiction but in the sense of what is generally accepted in the social context.

None of this, of course, would change the fact that the overall resources that a society has at its disposal are limited, nor would it change the fact that these resources have to be divided among its various domains of obligation. However, it does mean that an ethical society has an obligation to use the powers at its disposal to ensure that the limited resources that it can assign to the domain of health care will not be subject to cost decisions that are made solely by the corporations who provide the products.[40]

5.3 Conclusion

The discussion in this chapter has attempted to deal with the question of how an ethical society should structure the health care system that would provide the services that should be included in such a system, and some of the ethical implications that this would have. It did not start with the unargued assumption that health care services should not be considered commercial commodities but gave reasons for accepting this claim. It also suggested that in order for society to satisfy the obligation to provide equitable access to health care for its members. it would not only have to provide certain types of services—some of the details of this were already considered in the previous chapters—but that it would also have to structure access and availability conditions appropriately and effectively. Moreover, it suggested that this would involve not only material factors but also factors such as the education and training of a sufficient number of qualified health care professionals. |It further considered some of the related issues that would arise when engaging in the relevant activities: issues that include access to patient records for research and related purposes,

[38] *Encyclopedia of Management*, Pennsylvania State University, Gale, 2009, p. 31; Weisman (2002).

[39] World Trade Organization, *Agreement on Trade-Related Aspects of Intellectual Property Rights* (TRIPPS), accessed 28/11/2020 at https://www.wto.org/english/docs_e/legal_e/27-trips.pdf.

[40] For more on this, see the discussion in the Appendix.

as well as the use of appropriate technologically assisted modalities such as eHealth and telemedicine.

Much of what was presented was dealt with in only a cursory fashion. So, for instance, the implications for health care professionals and institutions were essentially discussed only insofar as their role in the framework of a social system of health care was concerned, whereas of course even in this context the ethical implications of their relationship to patients deserve more detailed consideration. Likewise, the ethical implications that a socialised system of health care has for the members of society themselves warrant more detailed discussion than what was said relative to the issue of social existence and safety, and a full treatment of this requires consideration of possible duties on part of patients themselves For, just like any other duty, the fact that society has an obligation to provide equitable health care to its members does not come without conditions on the right-holders themselves. In particular, it has implications for how members of society should structure their own health-related conduct. More specifically, just as an ethical society has an obligation to provide equitable health services for its members, so members of society have an obligation not to undermine an ethical society's ability to successfully carry out its duty by their engaging in inappropriate health related actions. These and related matters were only hinted at, and deserve further discussion. The chapters that follow are intended to fill this gap.

References

Aakvika, A., T.H. Holmås, and E. Kjerstad. 2015. Prioritization and the elusive effect on welfare—A Norwegian health care reform revisited. *Social Science & Medicine* 128: 290–300.

Ahn, D.G., H.J. Shin, M.H. Kim, et al. 2020. Current Status of Epidemiology, Diagnosis, Therapeutics, and Vaccines for Novel Coronavirus Disease 2019 (COVID-19). *Journal of Microbiology and Biotechnology* 30 (3): 313–324.

Baicker, Taubman, Allen, Bernstein et al. 2013. The Oregon experiment—Effects of medicaid on clinical outcomes. The New England Journal of Medicine, 368.

Bartz, C.C. 2016. Les soins infirmiers en télémédecine et télésanté dans le monde [Nursing care in telemedicine and telehealth across the world]. *Soins* 61 (810): 57–59.

Choi, P.J., R.J. Oskouian, and R.S. Tubbs. 2018. Telesurgery: Past, present, and future. *Cureus Journal of Medical Science* 10 (5): e2716.

Deepika, C., A. Kandaswamy, and C. Vimal. 2010. Protection of patient identity and privacy using vascular biometrics. *International Journal of Security* 4 (5): 64–84.

Doan, R. 2011. The false claims act and the eroding scienter in healthcare fraud litigation. *Annals of Health Law* Winter; 20 (1): 49–76, 8 p preceding 1.

Duckett, S.J. 2005. Private care and public waiting. *Australian Health Review* 29 (1): 87–93.

Ekeland, A.G., A. Bowes, and S. Flottorp. 2010. Effectiveness of telemedicine: A systematic review of reviews. *International Journal of Medical Informatics* 79 (11): 736–771.

Embrett, M., and G.E. Randall. 2018. Physician perspectives on Choosing Wisely Canada as an approach to reduce unnecessary medical care: a qualitative study. *Health Research Policy and Systems* 16 (1): 95.

Erkoreka, A. 2010. The Spanish influenza pandemic in occidental Europe (1918–1920) and victim age. *Influenza and Other Respiratory Viruses* 4 (2): 81–89.

Ewing, T.E. 2017. Will it come here? Using digital humanities tools to explore medical understanding during the Russian Flu Epidemic, 1889–90. *Medical History* 61 (3): 474–477.

Faunce, T., G. Urbas, L. Skillen, and M. Smith. 2010. Recovering fraudulent claims for Australian federal expenditure on pharmaceuticals and medical devices. Journal of Law and Medicine 18 (2): 302–15.

Fromer, M.J. 2006. FDA introduces new Phase 0 for clinical trials. *Oncology Times* 28 (15): 18–19.

Fuchs, V. 2009. Eliminating 'waste' in health care. *The Journal of the American Medical Association* 302 (22): 2481–2482.

Greenbaum, D., A. Sboner, X.J. Mu, and M. Gerstein. Genomics and privacy: Implications of the new reality of closed data for the field. *PLOS Computational Biology* 7 (12): e1002278.

Huang, Q., X. Yang, and S. Li. 2011. Identity authentication and context privacy preservation in wireless health monitoring system. *I.J. Computer Network and Information Security* 4: 53–60.

Kluge, E.-H. 2000. Medical narratives and patient analogs: The ethical implications of electronic patient records. *Methods of Information in Medicine* 38 (4–5): 253–259.

Kluge, E.-H. 2020. *The Electronic Health Record: Ethical Considerations*, Chapter 3.

Lasser, K.E., et al. 2006. Access to care, health status, and health disparities in the United States and Canada: Results of a cross-national population-based survey. *American Journal of Public Health* 96 (7): 1300–1307.

Legotlo, T.G., Mutezo, A. 2018. Understanding the types of fraud in claims to South African medical schemes. *South African Medical Journal* 108 (4): 299–303.

Manji, R.A., E. Jacobsohn, H.P. Grocott, and A.H. Menkis. 2013. Longer in-hospital wait times do not result in worse outcomes for patients requiring urgent coronary artery bypass graft surgery. *Hospital Practice* 41 (3): 15–22.

Miller, F.G., and H. Brody. 2003. A critique of clinical equipoise: Therapeutic misconceptions in the ethics of clinical trials. *Hastings Center Report* 33 (3): 19–28.

Newton, E., A. Janjua, E. Lai, G. Liu, T. Crump, and J.M. Sutherland. 2017. The impact of surgical wait time on patient reported outcomes in sinus surgery for chronic rhinosinusitis. *International Forum of Allergy & Rhinology* 7 (12): 1156–1161.

Nikolova, S., M. Harrison, and M. Sutton. 2015. The impact of waiting time on health gains from surgery: Evidence from a national patient-reported outcome dataset: waiting time and surgery outcomes. *Health Economics* 25 (8): 955–968.

Popovich, L. 2011. Russian Federation: Health system review. *Health Systems in Transition* 13 (7): 1–190, xiii–xiv. https://www.euro.who.int/__data/assets/pdf_file/0006/157092/HiT-Russia_EN_web-with-links.pdf. Accessed 03 Sep 2020.

Reichert, A., and R. Jacobs. 2018. The impact of waiting time on patient outcomes: Evidence from early intervention in psychosis services in England. *Health Economics* 27 (11): 1772–1787.

Sari, I., Z. Acar, Z. Nurkalem, N. Uslu, V. Davutoglu, M. Ates, O. Ozer, M. Eren, and M. Aksoy. 2007. Preoperative clinical status but not waiting time predicts in-hospital outcomes of surgery in patients with left main coronary artery stenosis. *The Tohoku Journal of Experimental Medicine* 213: 2, 173–180.

Sommers, Baicker and Epstein. 2012. Mortality and access to care among adults after state Medicaid expansions. *The New England Journal of Medicine* 367: 1025–1034.

Van Doorslaer, Masseria, Koolman. 2006. For the OECD Health Equity Research Group "Inequalities in access to medical care by income in developed countries". *CMAJ* 174 (2). https://doi.org/10.1503/cmaj.050584.

van den Heuvel, J.F., T.K. Groenhof, J.H. Veerbeek, et al. eHealth as the next-generation perinatal care: An overview of the literature. *Journal of Medical Internet Research* 20 (6): e202.

Weinstein, R.S., E.A. Krupinski, and C.R. Doarn. 2018. Clinical examination component of telemedicine, telehealth, mHealth, and connected health medical practices. *Medical Clinics of North America* 102 (3): 533–544.

Weisman, D. 2002. The law and economics of price floors in regulated industries. *The Antitrust Bulletin* 47 (1): 107–131.

Whitten, P., and F. Mair. 2004. Telesurgery versus telemedicine in surgery–An overview. *Surgical Technology International* 12: 68–72.

Woolhandler, S., T. Campbell, and D.U. Himmelstein. 2004. Health care administration in the United States and Canada: Micromanagement, macro costs. *International Journal of Health Services* 34 (1): 65–78.

Yip, J., W. Hao, A. Eskander, and J.M. Lee. 2019. Wait times for endoscopic sinus surgery influence patient-reported outcome measures in patients with chronic rhinosinusitis who fulfill appropriateness criteria. *Int Forum Allergy Rhinol.* 9 (4): 396–401.

Chapter 6
Health Care Professionals

Abstract The present chapter considers the implications that a society's acknowledgement of a right to health care has for the professionals who provide that care. In particular, it focuses on such issues as the nature of the fiduciary relationship that exists between the professionals and their patients, on informed consent, records, confidentiality and remuneration. Withdrawal of professional services is also considered, as are models of the professional-patient relationship itself.

Keywords Consent · Fiduciary relationship · Records · Remuneration · Withdrawal of services

6.1 Introduction

The analysis in the preceding chapters suggested that irrespective of legal, traditional or other considerations members of a society have an ethically grounded right to health care, and that an ethical society has a corresponding duty to provide such care for its members. It further outlined how the framework of appropriate health care services might be structured at the policy level in light of the fact that societal resources are always limited and that consequently, for pragmatic as well as ethical reasons, society has to apportion its resources among the different domains of services that it is obligated to provide. Chapter 4 sketched some considerations that would be ethically relevant at the hands-on level if the right to health care were to be realized, and outlined how this framework might be implemented in actual terms. Particular attention was paid to considerations that would be involved in resolving conflicting right-claims. Chapter 5 then dealt with some of the ethical issues that would arise when the obligation to provide an equitable system of health care is realized, with particular focus on the funding, distribution and accessibility of the relevant services as well as on matters such as the availability of the professionals who would provide the services and some of the technological implications and privacy issues that would arise in the context of delivery and planning.

One of the major points that was made but that was not fully addressed was that providing health care involves hands-on activities, and that because society does not itself act in a hands-on fashion its obligation to provide health care can be fulfilled

E.-H. W. Kluge, *The Right to Health Care: Ethical Considerations*, The International Library of Bioethics 92, https://doi.org/10.1007/978-3-030-93838-3_6

only through individuals who act, so to speak, as its agents. In real terms, the latter fall into two general groups: the professionals who provide the relevant hands-on services, and the institutions in which these individuals frequently exercise their profession. The focus of the present chapter is the professionals who actually deliver the health care in a hands-on fashion. The role of institutions will be the subject of the next chapter.

6.2 General Considerations

As was said a moment ago—and this bears repeating because it sets the framework for the ethics that underlies what follows—when health care is construed as a right, the individuals who provide that care at the hands-on level are individuals who actualize society's obligation to provide equitable access to health care. For that reason, and purely ethically speaking, as long as they function in that role, what they do and how they do it is subject to the same overarching ethical principles that ground society's duty to provide health care. This entails that irrespective of the particular nature of the health care activities in which they engage, there are certain general and ethically grounded conditions that should structure the relationship that they have with their patients.

The reason this holds true irrespective of their specialty and irrespective of the particular nature of their professional activities is that the particulars of what they do—more precisely, perhaps, the specific interventions in which they engage when these are considered in and by themselves—are, so to speak, a material matter whereas the framework within which they operate is an operational matter. Thus, considered in and by itself and in a material sense, providing a narcotic or an analgesic is ethically neutral, as is providing insulin, filing a prescription or performing a surgical operation. It is the context in which these actions are performed that is ethically relevant. Therefore, unlike the purely material aspects of their activities, it is what was referred to as their operational aspect that makes these activities subject to ethical constraints.

This is not simply a theoretical consideration. It is also reflected by the fact that there is a fundamental difference between codes of professional conduct or ethical considerations on the one hand, and technical protocols and guidelines for best practices on the other.[1] The former stipulate how professional conduct should be ethically structured, the latter identify the practical details of what to do when faced with a specific problem and what action is most likely to produce the best results

[1] Cf. Meriam-Webster Dictionary, "best practice", accessed 14/08/2020 at https://www.merriam-webster.com/dictionary/best%20practice. See also Cambridge Dictionary at https://dictionary.cambridge.org/dictionary/english/best-practice.

in the most efficient way on a given occasion, as this is sometimes outlined in such things as ISO standards.[2]

6.3 Health Care Professional as Society's Agents

This brings into focus the previous claim that irrespective of the particular nature of their professional expertise, the health care professionals who engage in providing health care that falls within the realm of services that an ethical society is obligated to provide for its members effectively function as society's agents.

At first glance, this claim may seem absurd. At best—so the counterargument might run—this holds true only in those jurisdictions where health care is a purely socialized affair, is delivered only by governmentally funded and controlled institutions, and where all health care professionals are employed by these institutions—in a word, if they essentially are government employees.

Moreover—so the counterargument might continue—that is neither universally the case nor does it really reflect the situation even in those jurisdictions where governments take an active role in providing healthcare: for instance in countries such as Albania, the Peoples' Republic of China and Canada. Even in these countries, health care professionals are free agents and they may practice their professions outside of and independently of any settings that are socially financed or controlled. Moreover, health care professionals generally do not consider themselves to be society's agents. They see themselves as independent practitioners, and point to the fact that they are generally paid either directly or indirectly by their patients. Therefore to portray health care professionals as society's agents is to seriously misrepresent what is the case.

However, the intuitive appeal of these considerations notwithstanding, they miss the point of what is at issue. In the first place, how someone sees their role is not necessarily an indication of the correct way of assessing that role. So, for instance, parents may perceive themselves as having the right to make decisions for their young children because they see themselves as owning their children. In fact, however, such a perception is mistaken. Parents do not own their children but only function as their children's caretakers and as their substitute decision makers. Moreover, they may do this only as long as they make appropriate decisions and the children lack competence.[3]

Secondly, what health care professionals do within the parameters that were indicated—i.e., providing the kinds of services that an ethical society is obligated to provide—is different from what they do outside of that context when they provide health services that do not fall within this domain. However, as soon as they provide services that fall within the former domain their practice is ethically constrained

[2] International Organization for Standardization, *Standards*; accessed 15/10/2020 at https://www.iso.org/publication/PUB100343.html.

[3] Bester and Kodish (2017), Montgomery (1988).

by the principles that structure society's obligation to provide health care for its members. This holds true even when they practice their profession as independent practitioners; and it holds true irrespective of the method of their recompense because that affects only the manner of remuneration, not the nature of what they do.

Because this point is important, it may be worth exploring it a bit further. There is a logical difference between what one does, why one does it and how one is recompensed for whatever it is that one is doing on the one hand, and the overarching conditions that apply to what one is doing on the other. So, for instance, a lawyer may provide legal services to a client, and while the lawyer may be paid by the client, when they are representing their client's interests in a court of law they in fact function as officers of the court, and their actions are subject to the conditions that are built into the legal framework that structures how laws are to be interpreted and applied.[4] In that sense, they are society's agents even though they represent their clients and are paid by their clients and not by society. In other words, their role as professionals is to ensure that the legal provisions that society has enacted are applied to their clients according to the rules that society has put in place. The fact that their reason for acting as lawyers is their clients' desire to have their rights enunciated and defended, or that their actions are ultimately grounded in their own desire to be recompensed, is an entirely different matter.

Or, to bring the matter closer to home, the fact that health care services of the sort that were identified in Chap. 3 are provided by a privately funded charitable organization in a society that recognizes health care as a right—for instance, by the Aga Khan Health Services[5]—does not entail either that the services that it provides in that jurisdiction are commodities or that the physicians who provide these services are not acting as society's agents. To be sure, they are acting at the instigation of the Aga Khan Health Services and in that sense they are also acting as agents of the Aga Khan Health Services, but that has to be seen in the overarching context of the health care system of the jurisdiction where they provide their services. The fact that the Aga Khan Health Services takes over the cost of providing these services rather than leaving it to the jurisdiction (or to individual patients) to defray does not change the fact that the physicians are professionally moving within the framework that ethically governs the society's health care delivery system. The Aga Khan Health Services may even expand the types of services that are delivered beyond what would otherwise be available in that social setting. However, the fact remains that the health care professionals whom it hires would still be working under the overarching ethical framework that governs (or that should govern) the society's obligation to provide health care on an equitable basis. Expanding the domain of what is delivered merely means that the Aga Khan Health Services are enabling the society to fulfil its obligation in a way that the society would otherwise be unable to do. Hence, once more, the professionals are *de facto* functioning as social agents when they act in their professional capacity.

---

[4] Goethe (1989).

[5] https://www.akdn.org/our-agencies/aga-khan-health-services/about-aga-khan-health-services.

With due alteration of detail, similar considerations apply to the physicians who work under the helm of an association like *Médecins Sans Frontières*. While the physicians who exercise their profession within the administrative context of that association are paid by the association,[6] the services that they provide fall within the framework of health care services that an ethical society is obligated to provide for its members. Therefore, ethically, they too function professionally as agents of such a society. It is merely that the burden of having to pay for their services is not shouldered by that society, and that the society's ability to satisfy its ethical obligation is thereby expanded.

It should be noted that while the discussion so far has focused on physicians, with due alteration of detail analogous considerations apply to health care professionals in general. In other words, they also apply to individuals such as medical assistants, nurses, physiotherapists, dentists, midwives, psychologists, psychiatrists, or pharmacist if they deliver the kinds of services that fall under the rubric of socially obligated health care services in a rights-oriented society as was sketched above.

Moreover, the limiting clause that was just mentioned—namely, "the kinds of services that fall under the rubric of socially obligated health care services"—is crucial and should not be overlooked in this connection. Not everything that health care professionals do falls within the domain of health services that an ethical society is obligated to provide. So, for instance, non-restorative cosmetic surgery is not something that would fall into this domain, and neither would medical interventions that are intended to improve or even maximize the athletic abilities of amateur or professional athletes.

6.4 Models of the Professional-Patient Relationship

The fact that health care professionals function as society's agents when providing the kinds of health care that society is obligated to provide for its members raises the question how such professionals should be positioned relative to patients when they provide these services. More colloquially, it raises the question what model of the professional-patient relationship they should adopt. The reason this question is important is that there is a variety of different approaches, and they entail different things with respect to how—or indeed, whether—patients have access to these services.

For instance, the kind of approach where the professional adopts what is essentially an authoritarian approach—what is sometimes called a paternalistic or priestly model[7]—allows the professional to decide what services are appropriate and therefore which of the socially guaranteed services would be appropriate in a given case

[6] https://www.doctorswithoutborders.ca/terms-employment.

[7] For classic discussions of model of the physician–patient relationship see Veatch (1972), as well as Childress and Siegler (1984), Emanuel (1991). For a somewhat mitigated position on patient autonomy, see Tauber (2005), and Childress and Siegler (1984).

and what would be in the patient's best interest—where the criterion of what would be in the patient's best interest would be what is considered by the professional to be appropriate. On this model, the values that are espoused by the patient essentially play no role.

Alternatively, on what is sometimes called a collegial or partnership model, the professional and the patient collaborate as equals in deciding what would be appropriate, and their interaction proceeds on that basis. By contrast, on a contractor or engineering model the patient selects which of the services they would want and the professional merely indicates whether (and why) the services would be professionally appropriate and whether the physician will act on the patient's decision. In effect, then, the professional then merely has a service to offer and hands over the decision-making to the patient. By contrast, on a friendship model the professional acts towards the patient the way a friend would, and like a friend tries to persuade the patient to accept or reject a particular kind of intervention. On this model, there is a personal relationship between the professional and the patient, and it plays an important role in their interaction. The professional may even try to pressure the patient in a particular direction, just as a friend would on occasion.

Finally, the fiduciary model involves a trust-based relationship between the professional and the patient. The greater knowledge and expertise of the professional is fully acknowledged, but so is the right of the patient to ultimately decide whether to accept or reject a given intervention. While the professional will always act in the best interest of the patient, in contrast to the paternalistic model what counts as being in the patient's best interest is determined not on the basis of purely professional criteria but in accordance with the patient's values.

Ethically speaking—and tradition aside—the model that best accords with health care considered as a socially guaranteed right and with the role of health care professional as functioning as society's agent is the fiduciary model. There are several reasons for this. The first and perhaps most important is that if the analysis in the preceding chapters is correct, then health care is a right that belongs to members of society. This means that all other things being equal—the matter will be explored further in Chap. 8—it is not a duty, which in turn means that like all other rights it may be exercised at the discretion of the right-holder as based on their values. At the same time, the professional possesses the knowledge and the training—knowledge and training that the patient usually lacks—that allows them to evaluate what intervention (if any) is medically appropriate for the patient's condition. Moreover, in most contemporary jurisdictions health care professionals are the only ones who are legally permitted to provide the services in question. All of this means that patients must be able to trust that the health care professionals will act in their best interest when they make suggestions, and that the professionals will exert their best efforts when engaging in the actions to which the patients have given consent.

6.5 Withholding or Withdrawing of Services

One of the more important considerations that were raised in the preceding discussion, albeit only in passing, was that in most jurisdictions licenced health care professionals are the only ones who are legally permitted to engage in the kinds of actions that fall under the rubric of medical services. This puts health care professionals into a special position not only in the sense that they effectively function as society's agents but also in the sense that the services which fall under the rubric of socially mandated health care will not be available if the health care professionals choose not to provide them. This raises the question whether, in a rights-oriented health care system, health care professionals have the right to make such a choice.

To put this into context, it may be useful to recall that the Principle of Autonomy entails that everyone has the right to self-determination subject only to the equal and competing right of others.[8] At first glance, this not only means that health care professionals have the right to choose their particular profession but also that they have the right to withhold or withdraw their services if they so wish.

However, there are several reasons why this would not be ethically appropriate in a rights-oriented health care system. Thus, if all and only those individuals who have been trained in a particular health care profession and have passed the qualifying examinations that attest to their competence will receive a license to practice—an overarching condition that is usually reflected in binding legal provisions—then if the members of that profession were to withhold or withdraw their services, no-one else would legally be authorised to step in and take their place. In that case, however, the health care that would otherwise be provided by the members of that profession would be unavailable, which in turn would mean that the right to health care that belongs to the members of society could not be satisfied.

Arguably, therefore, this suggests that there is a clear limit to the right of self-determination that otherwise belongs to members of a health care profession if health care is construed as a right. More specifically, it suggests that the fact that health care professions are legally sanctioned service provider monopolies essentially entails that health care professionals may not ethically withhold or withdraw their services.

But even here it is arguable that there are limits. While it does mean that health care professionals may not withhold or withdraw from a case because it poses a risk to the professional themselves—the matter will be discussed in a moment—it does not mean that professionals are ethically obligated to accept cases or otherwise engage in providing services if doing so would transcend their competence or, in case of a relationship that has already been established, if it would result in errors or mistakes in providing the relevant service. Thus, requiring the professionals to work longer than a normal span of time would be ethically inappropriate because it would significantly increase their risk of error,[9] which would be at variance with their duty to always act in the best interest of their patients. It is only if no other option was available and the necessity condition was satisfied, that the professionals would have

---

[8] Cf. Chap. 2, p. 22.

[9] Caldwell et al. (2019), Carrez et al. (2019), Pfaff (2004), Tang et al. (2019).

a duty to continue to act to the best of their ability—however, only on the condition of having apprised their patients (as well as appropriately placed other parties) that this would be the case.

However, once more there are certain conditions that qualify what has just been suggested. More specifically, it does not mean that health care professionals are obligated to work as health care professionals if the conditions under which they are expected to work or the tools with which they are provided are inappropriate, ineffective or otherwise unsuitable. All it means is that by choosing to become a member of a certified health care profession, they effectively accept the condition that they will adjust their autonomous actions as professionals in keeping with the overarching condition that they and no-one else may legally work in that area.

6.6 Implications of the Fiduciary Relationship

To clarify this further, it becomes necessary to once more expand the discussion and return to another overarching factor that conditions the ethical nature of the framework within which health care professionals function in a rights-oriented health care system. As was previously pointed out, health care professionals stand in a fiduciary relationship to their patients. This relationship may be incepted directly when they interact with patients without intermediaries—as for instance when they have direct physical contact with them or interact with them directly by electronic means as would be the case in direct virtual care or direct eHealth and telemedicine—or it may be initiated indirectly, as when they provide professional services on a consultative basis and provide services that will be used by another health care professional who stands in a directly incepted relationship with the patients.[10]

Because they stand in a fiduciary relationship to their patients and in the context of a rights-oriented health care system, health care professionals have an obligation to act in the latter's best interest. Applying this to the issue at stake it means, among other things, that ethically speaking they may not withhold or withdraw their services if the patients depend on those services. If they wish to do so, or if they have reasons for doing so, they have a duty to inform their patients in due time of this fact[11] and try to find an appropriately qualified substitute who would be available in their place, or to offer the patients the opportunity to seek out another appropriately qualified professional.[12] If the patients did not wish to do so, or if time was of the essence and the type of intervention that was involved satisfied the criterion of necessity that was discussed in Chap. 4 and, moreover, if no other duly qualified professional was available to step in, then they would have a duty to inform their patients (or significant others) of the situation and continue to provide whatever service they

[10] Cf. Chap. 5, p. 96.

[11] Or the patients' duly empowered substitute decision makers, as the case may be.

[12] Cf. World Medical Association *International Code of Medical Ethics*, available at https://www.wma.net/policies-post/wma-international-code-of-medical-ethics/.

could. Anything less would constitute abandonment of the patient and a violation of their fiduciary obligation.

As to the type of situation where interacting with a patient would pose a serious health-risk to the professional—highly infectious diseases such as Ebola or COVID-19, etc. here come to mind—this would not constitute an ethically valid exception to the duty to provide the implicated services, for here the fact that entering a licenced health care profession and the fact that they would be practicing in a right-oriented health care system limits their autonomy once again comes into play. By choosing to enter a profession that is a service provider monopoly, they bind themselves to provide the relevant services that are the mandate of that profession.[13] The health risk that this may sometimes pose to their own person is something that goes with the initial choice of the profession itself, just as much as accepting the risk of being injured goes with choosing to become a firefighter, a soldier or a police officer. And to assume that some other member of the profession would step in their place and would deliver the relevant services if they decided to withhold or withdraw them would be to assume that other similarly placed professionals would not make the same decision—which would be an unwarranted assumption.

Finally, and looking at the larger picture, it would be to invalidate society's reason for establishing health care professions as services provider monopolies in the first place. The fact is that an ethical society does not establish service provider monopolies for the benefit of the monopoly holders if the type of service at issue involves a socially guaranteed right. The reason for establishing it as a service provider monopoly is to ensure that the members of society will have equitable access to the relevant service at an appropriate time, of appropriate quality and at an appropriate level of expertise. This is distinct from society establishing service provider monopolies where the services that are delivered do not involve socially guaranteed rights but involve trade or such things as intellectual property, etc. Socially guaranteed health care services are not commodities in that sense and are not patentable either by law[14] or by ethics.[15] While providing health care services may involve the use of patentable items such as those of a technological nature, that is another matter.

At the same time, one of the assumptions that underlies the inception of a professional-patient relationship—and this is certainly the case in an institutional context such as a hospital or a "hospital in the home" in some countries such as Australia[16]—is that the conditions under which the relevant services would be performed would meet otherwise appropriate professional standards. This includes time parameters as well as such things as equipment, tools and even assistance as

[13] Sim and Purtilo (1991), Yakubu et al. (2016).

[14] World Trade Organization, *Agreement on Trade-Related Aspects of Intellectual Property Rights*, s. 27:2–3; accessed 28/11/2020 at trips_art27_jur.pdf (wto.org). For further discussion, see Chap. 7, *infra*.

[15] World Medical Association, *WMA Statement on Patenting Medical Procedures*; accessed 21/10/2020 at https://www.wma.net/policies-post/wma-statement-on-patenting-medical-procedures/.

[16] https://www2.health.vic.gov.au/hospitals-and-health-services/patient-care/acute-care/hospital-in-the-home.

and when this would be necessary. When it is clear that this assumption would not be met, it would normally be ethically appropriate for a health care professional to decline entering that setting. However, if the only way in which a patient's right to health care could be satisfied was if the professional acted in this sort of setting—as would be the case in some outlying communities—it would be ethically mandatory for the professional to function professionally in such a setting but it would also be mandatory for them to inform their patients of the issues that trying to provide the services under such conditions might present.

6.7 Conscientious Objection

Of course health care professionals are also persons and therefore, like all other persons, have the right not to have their basic values violated. However, there may be situations where a patient's request or need for a particular kind of health service may otherwise be ethically defensible and may even be one of the services that is recognized as falling under the right to health care—abortion or medical assistance in dying may here be implicated—but providing the particular kind of service would go against the individual professional's fundamental values. It is generally accepted that under such circumstances health care professionals have a right to refuse such a service, the fiduciary nature of the relationship with their patients notwithstanding, because they have a right not to have their own values violated. At the same time, it is also generally accepted that under such circumstances they should refer their patients to another professional who would not be thus constrained by their own personal values.[17]

However, while this latter position has legal approval in some jurisdictions,[18] that is solely a matter of law and does not settle the matter of ethics; and as one survey of physicians has shown, physicians themselves are divided about a professional obligation to refer when an individual physician believes that referral itself is immoral.[19] Thus, some physicians contend that by referring their patients to another professional who will perform the service that they themselves find morally objectionable they would become co-responsible for the act, and therefore that referring their patients would not absolve them of moral responsibility. Therefore, so they argue, in such cases the duty to refer would be tantamount to the duty to become ethically guilty.

At first glance, therefore, the issue of conscientious objection presents such health care professionals with an insoluble dilemma. On the one hand, it would violate their fiduciary obligation if they did not refer when a referral was otherwise professionally appropriate; on the other hand they would become morally guilty according to the ethical standards that they espouse if they did refer.

---

[17] Ancell and Sinnott-Armstrong (2017), Cowley (2017).

[18] *Christian Medical and Dental Society of Canada v. College of Physicians and Surgeons of Ontario*, 2019 ONCA 393.

[19] Combs et al. (2011).

However, to characterise this sort of situation as a dilemma is to misrepresent it, because the issue itself arises at a much earlier stage than when the conflict-situation actually presents itself in a rights-oriented health care system. To put it first in general terms, while the Principle of Autonomy entails that an individual has the right to choose a particular profession, the exercise of that right is rationally and ethically valid only if the individual understands the nature of what they are choosing. When this consideration is applied to the choice of a health care profession in this particular context, then when someone considers becoming a health care professional in this kind of setting they should be aware that while the potential for such a conflict situation does not necessarily arise in all health care professions—for instance, it would not usually arise in physiotherapy or dentistry—that type of conflict is not unheard of in such health care professions as medicine, nursing or pharmacology.

Consequently, when entering a health care profession that is embedded in a society that considers equitable access to health care as a socially guaranteed right and that type of health care profession involves providing certain types of services that the individual would find personally objectionable, then the individual who chooses to become that type of health care professional in such a setting would be choosing to do so in full awareness of the fact that they would be committing themselves as a matter of fiduciary obligation to providing such services either directly themselves or indirectly by referring their patients to someone who will provide them when they themselves would find the intervention morally objectionable.[20] Therefore, the dilemma that was outlined above should never arise at the hands-on level in the first place. A reasonable and ethically conscientious individual would never choose to enter a health care profession where such a dilemma would arise.

By contrast, when care professions are not exercised in the context of a socialized health care system—in other words, when heath care is considered a commodity—the situation would be fundamentally different simply because then health care services would be commodities that the professionals would have for sale and that the patients could purchase. Therefore the health care professionals could simply refuse to sell any services that they found personally objectionable. Of course if they did sell a particular service, then they would be duty bound to provide it to the best of their ability, and their personal scruples could no longer enter the picture. At the same time, however, they would have the right to stipulate at the beginning of the professional-patient interaction that they would not provide the relevant type of service even if such a service was otherwise available from other health care professionals, or even if such a service was included in the kinds of services that are recognized as legitimate within the overall domain of health care or in that society. Nor would they have an obligation to refer their patients to other health care professionals who performed the relevant intervention, any more than any other vendor of a commodity would have an obligation to refer a customer to another purveyor if they found the sale of such a commodity to be morally objectionable according to their own standards. At the same time, however, the possibility of such a refusal should be clearly explained to possible patients before the inception of an actual professional-patient relationship.

---

[20] Cf. Savulescu (2006).

6.8 The Professional as Infectious Agent

A further ethically sensitive issue that falls under the rubric of withholding or withdrawing services—and one that has particular relevance at the time of a pandemic—is whether the fiduciary obligation of health care professionals either requires or allows them to continue to interact with patients when they themselves are infected with a communicable disease. The matter assumes particular importance when the health care system within which they operate has only limited capacity, and that for them to withhold or withdraw their services would mean that the system would be understaffed and patients would not receive the services that they require and to which they are entitled as a matter of right.

One again, this appears to face health care professionals in a rights-oriented system with a dilemma: If they withheld or withdrew their services—which apparently would be in keeping with the Principle of Nonmaleficence—they would not run the chance of infecting their patients and thus not run the risk of causing harm. On the other hand, if they withheld or withdrew their services in times of professional scarcity they would still cause harm because the otherwise necessary services would not be available. Apparently, therefore, they would again violate the Principle of Nonmaleficence, albeit in a different way.

Once again, however, it is arguable that the dilemma is more apparent than real. In the first instance, this sort of situation would pose a dilemma if and only if there was a staffing shortage. If that was not the case, then the Principle of Nonmaleficence would entail that the professionals should inform their patients of their own infectious status and they should normally refrain from any interactions that would have a reasonable chance of being contagious. At the same time, however, the chance of contagion would not necessarily exist in all cases. For instance, no danger of transmission would exist if the professional-patient interactions were virtual in nature, such as in eHealth and telemedicine, nor would it be the case when only distanced and indirect laboratory procedures or similar undertakings would be involved.

Moreover—and this brings the issue back to the qualification that was expressed above by saying that this should normally be the case—if the professionals informed their patients of their own status and clearly explained the possibility of transmission and if, moreover, the patients already suffered from the same condition and agreed to the continuation of the direct professional-patient interaction, then the Principle of Nonmaleficence would not be violated if the professionals did not withhold or withdraw their services. No additional harm would ensue because the patients would already be infected and, moreover, the patients' consent would be sufficient and would be ethically acceptable because it would be an instance of the patients' exercise of autonomy.[21]

As to the sort of situation where staffing was limited and appropriate services could not be provided if the professionals were to withhold or withdraw their services, there would be two ethically appropriate ways of dealing with the situation. If the infection status of the professionals was like that of the patients, then the situation would be

[21] For more on the issue of informed consent, see *infra* pp. 119 ff.

similar to the sort of case that was discussed a moment ago. The transmission of the causal agent from the professionals to the patients would not cause any further harm beyond what already existed because the patients were already infected with the particular condition. Therefore, as in the previous case, the Principle of Nonmaleficence would not be violated.[22] Of course even in such cases the patients should be told of the infection status of the professionals and should give consent, because it is a general but understood assumption that health care professionals are not themselves carriers of disease processes, and that they should inform their patients if that is not the case.

The other ethically acceptable option also hinges on the issue of informed consent. To be precise, if the patients did not suffer from the same infection as the professionals, then clearly there would exist the possibility that even if the professionals were to take the best possible preventive measures, the infection could be passed on to the patients. However, the Principle of Autonomy and the ethics of informed consent would here enter the picture. If the intervention in question would be an intervention that would satisfy the patients' health care needs, and if that health care need was equivalent to or greater than the health care need that the patients would experience if the patients were to become infected then, if the patients understood the nature of the situation and gave informed consent to interacting with the professionals, the professionals would not need to withhold or withdraw their services. On the other hand, if the health care needs that would be generated if the patients were to become infected would be greater than the health care needs that would not be satisfied if the professionals were not to provide their services, then the professionals would be ethically obliged to withhold or withdraw their services, the fact that the patients would then not receive the otherwise appropriate services notwithstanding. While it would be tragic that the patients' needs would then not be met, not everything that is tragic is unethical. Nor would it matter if the patients were not already infected but were to give consent. The Principle of Nonmaleficence would be engaged by the fiduciary nature of the professional-patient relationship and would not be overruled by the patients' consent. Therefore such consent would not impose an obligation on the professionals to become unethical in their actions.

At the same time, it is important to note that the preceding considerations focused on the type of situation where health care is construed as a socially guarantee right. The situation would once more be different if health care was construed as a commodity. While traditionally the same sort of considerations that were raised in the rights-context would be transferred to the commodity context and an appeal would be made to the fiduciary relationship that holds between health care professionals and patients, that would be a matter of tradition and not a matter of logic. Since, logically speaking, on a commodity model the interrelationship between health care professional and patient would be a matter of contract, strictly speaking the issues would be addressed by the terms of the contract that obtained between the professionals and the patients. And while this would not mean that ethical considerations

[22] Clearly, this assumes that the infectious agent was of the same type, and that the patient had not recovered from the infection.

would entirely drop out of the picture, these would be considerations that would be derived from business ethics, not in health care ethics. Nor would they be grounded in considerations that derive from construing health care as a right. To argue otherwise would be to rely on tradition, not on logic.

6.9 Professional Associations

As emerged in the preceding discussion, the issue of informed consent plays an important role in the professional-patient interaction and as such merits further consideration. However, before doing so, a few words about associations of health care professionals would not be amiss because the latter frequently provide their members with guidelines regarding the withholding and withdrawal of health care services.

In this regard, it may be appropriate to begin by indicating a fundamental distinguishing feature of health care professional associations or organizations on the one hand, and associations or organizations of other kinds of professionals on the other.

Like other kinds of service providers, health care professionals sometimes form associations, organizations or societies whose purpose is to safeguard or even advance the interests of their members and to present a systematized picture of the nature of their profession to the rest of society. In fact, in most jurisdictions this is the case. These associations, organization or societies tend to be specific relative to membership criteria. Thus, in the domain of health care, there are medical associations, nurses' associations, therapists' associations, pharmacy associations, etc. However while, just like all other associations, organizations or societies, their ultimate focus centres in advancing the interest of their members, how they function and what they do in a rights- as opposed to a commodity-oriented system of health care is constrained by the ethical parameters that structure the social embedding of their members as health care professionals in a rights-oriented system. In other words, such association or organization cannot and do not function quite the same way that associations of car dealers or trade organizations function, because the members of the latter types of associations or organizations are not embedded in a rights-oriented framework, nor do their practitioners enter into a fiduciary relationship with their clients.

Among other things—and this is what makes the roles of such associations relevant when considering the withholding or withdrawal of professional services—they tend to provide guidelines and codes of conduct for their members in order to assist them in the conduct of their professional lives. These guidelines and codes deal with a variety of issues, ranging from how to treat patients[23] to how to behave in times of

[23] World Medical Association, *International Code of Medical Ethics*. accessed 15/11/2020 at https://www.wma.net/policies-post/wma-international-code-of-medical-ethics/.

armed conflict.[24] Consequently, given what was said a moment ago, this means that appropriately constructed codes or guidelines would indicate the conditions under which the withholding or withdrawal of professional services by their members would be ethically appropriate. This in turn means that, in keeping with the preceding considerations, they would include a clear statement to the effect that because the services in which their members as licenced professionals engage fall into the domain of services that society is ethically obligated to provide, what they provide constitute essential services—i.e., services that are vital for preserving the health and welfare of the population.

Consequently, an association of health care professionals in the context of a rights-oriented system of health would promulgate codes or guidelines that would enjoin their members to refrain from what may loosely be described as wholesale job action like complete withdrawal of services, work slow-downs or the like for such things as the level of remuneration that they receive for their services. They would therefore instruct their members that any such job-action should take place only within the limits that were indicated above, and that neither the association nor its members could support wholesale withholding or withdrawing of services or of strikes if that would curtail essential services.

At the same time, the reason for contemplating job action may not always be based on the interests of the professionals—at least not in a material sense There may be occasions where the conditions under which the professionals are expected to work are not in keeping with the standards that the professionals are ethically entitled to expect or, more particularly, that would violate the fundamental and principled ethical considerations that were outlined in the preceding chapters. Consequently, when appropriately developed, the codes and guidelines that they promulgate would enjoin their members that if the issue could not be resolved in any other way, then they have an ethical obligation to resolve the matter by means of binding arbitration that acknowledges and respects the distinct parameters in question.

As to situations where the legal provisions that are enacted in a given jurisdiction would require health care professionals to engage in acts that would violate the otherwise ethical obligations that health care professionals have in virtue of the nature of their profession—situations that would require participation in such things as torture or other violations of human rights—health care professional associations would be ethically bound to inveigh against such provisions and enjoin their members not to act in accordance with these, but to adhere to the fundamental principle of medical ethics.

---

[24] World Medical Association, *Regulations in Times of Armed Conflict and Other Situations of Violence*; accessed 15/11/2020 at https://www.wma.net/policies-post/wma-regulations-in-times-of-armed-conflict-and-other-situations-of-violence/. For more policies, see WMA Policies, https://www.wma.net/policy/.

6.10 The Fiduciary Health Care Professional-Patient Relationship

One of the more important and possibly more controversial claims that was made in the preceding sections is that construing the nature of the fiduciary professional-patient relationship in ethical terms that are functionally derived from the fundamental ethical principles that were outlined in the preceding chapters is ultimately a function of construing health care as a right rather than as a commodity. This seems to present a bit of a puzzle. It seems to fly in the face of the fact that understanding health care as a commodity has an ancient tradition, and the construal of health care as a right did not occur not until the later nineteenth century when Germany did so with the introduction of the *Health Insurance B*ill of 1883, the *Accident Insurance Bill* of 1884, and the *Old Age and Disability Insurance Bill* of 1889.[25] At the same time, however, construing the nature of health care professional-patient relationship in fiduciary terms seems to go back at least as far as Hippocrates and the Hippocratic Oath, which latter was written at some time between the fifth and third centuries CE[26] when health care was not construed as a right.

However, one of the more important facts that is generally overlooked in this connection is that the Hippocratic Oath itself was developed by what historically was a relatively new school or association of medical practitioners that surrounded Hippocrates, and that it was administered to individuals on entry into this association and was distinct from the school or association of medical practitioners who practiced under the influence of the more religiously oriented schools of medicine that also existed at that time.[27] Therefore it was much like the solemn undertaking that individuals in ancient time affirmed when they became soldiers,[28] gladiators[29] or entered other professions.[30] Moreover, even as time went on and the Oath (or versions of it) became an integral part of entering the medical profession, the fact remained that it was the expression of a commitment on part of individuals who entered the profession of medicine, and that it was not integral to the provision of health care as a right. Nor was it even part of the framework in which health care was delivered in ancient Greece—for instance, of the framework of those who engaged in surgery or who provided abortions—as the Hippocratic Oath itself stipulates. Furthermore, nothing like this was integral to the practice of health care in non-Western cultures, as for instance in Ayurvedic or First Nations health care. Finally, even in the Western world it was not until surgery became absorbed into medicine in the eighteenth and

---

[25] Carrin and James (2005).

[26] Edelstein (1943).

[27] Kleisiaris et al. (2014), Marketos and Poulakou-Rebelakou (1995); *World History Encyclopedia*, "Ancient Greek Medicine", accessed 28/04/2021 at https://www.worldhistory.org/Greek_Medicine/

[28] Nock (1952).

[29] Seneca, *Epistulae* 71.32.

[30] Mousourakis (2007).

nineteenth centuries[31] that the Oath and the sentiments expressed in it became part of health care professions in general.

Also, it is arguable that quite aside from all this, the recognition of the fiduciary characteristic of health care professions as such is a function of the fact that health care professionals are professionals: more specifically, it is a function of the role that professionals play as professionals and of how the notion of a profession is understood in general. Professions in general are fiduciary in nature because professionals, on the inception of their relationship as professionals, undertake to act for, on behalf of or towards their clients in a particular matter or in particular circumstances, where it is this undertaking that gives rise to a relationship of trust and confidence. And the fact that the Hippocratic Oath and its derivatives essentially include the promise to always act in the best interest of one's patients has ethical implications only on the assumption that it is ethically obligatory to keep one's promises. That, however, is not itself an ethical principle and—as philosophers have debated over the centuries—would have to be established on its own terms.[32] Moreover, it holds—if it holds—for all promising and is not unique to health care. Therefore it cannot be characterised as distinguishing health care ethics from any other domain.

Nevertheless, and for the reasons that have been outlined above, reference to the fiduciary nature of the relationship that holds between health care professionals and patients in the ethically embedded sense that includes reference to the fundamental principles of ethics in the way these function in health care, it holds only when health care is considered to be a right. Absent the overarching framework of health care as an ethical right, the fiduciary relationship that health care professionals have towards their patients is no different from the fiduciary relationship that all professionals have towards their clients, no matter what the professional specialty in question. While this may include the duty to act in the best interests of their clients and may involve ethical parameters, for the reasons that have also been pointed out above these are distinct from the ethical considerations that apply to health care professionals who practice their profession in the domain of health care construed as a right because they practice their profession in the unique ethical field that was outlined in the preceding chapters.

6.11 Informed Consent[33]

An issue that the analysis so far has passed over without much notice but that plays a tremendous role in the interactions between health care professionals and patients

[31] Aggarwal (2010).

[32] *Stanford Encyclopedia of Philosophy*, Promises, accessed 29/04/2021 at https://plato.stanford.edu/entries/promises/#ProProObl.

[33] For a discussion of the implications of different conceptions of personhood on the ethics of informed consent, see Chap. 7, pp. 143 ff.

is the issue of informed consent. It is particularly important when health care is construed as a right, and when society has a corresponding obligation to provide it.

The heart of the issue lies in the fact that all other things being equal—there are exceptions to this which will be addressed more fully in Chap. 8—health care is a right and not a duty. A right, of course, as was indicated in Chap. 1, is a justified claim. This means that the right-holder may exercise that claim but need not do so. Given, then, that health care is a right, this has two important consequences. The first is that having a right to health care is meaningless in practical terms unless the right-holder is aware of the existence of the right and of its extent and nature; and second—again, all other things being equal—that the right-holder has the option of refusing to exercise the right on any given occasion. In other words, that the term "consent" is actually short for "consent or refusal."

However—and this is what marks the matter as relevant in the present context—it also means that whoever engages in an activity that would be involved in actualising the right to health care on a given occasion has a corresponding duty to ensure that the right-holder—which is to say, that the patient (or the duly authorized representative of the patient)[34]—not only understands what is at issue but also makes an informed and competent decision about whether the intervention should take place.

The matter is sufficiently important to merit further discussion since what informed consent amounts to and how such consent is achieved is frequently misunderstood or even mishandled.[35]

6.12 Consent Itself

To begin at a very general level, consent in health care is the voluntary acceptance or rejection of a particular intervention by a particular patient. The disjunction itself is important. As was indicated before, it means that the term "consent" is really short for "consent or refusal."

A further important general point is that consent has an important cognitive component. This means that a consent is not ethically valid unless it is given on the basis of information that is sufficient in detail and clarity so that the individual who gives the consent not only can in principle understand the nature of the choice that they are making but in fact does understand it. This in turn means that in order to satisfy this criterion, the information must be offered in a way and at a level that is appropriate for the specific individual in question. A one-size-fits-all approach to providing the relevant information, therefore, is ethically inappropriate and comprehension at the level of the objective reasonable person is also insufficient. After all, it is not the objective reasonable person who is to give consent: it is the particular patient.

[34] For more on this, see *infra*, "Substitute Decision-making".

[35] Freedman (1975), Manson and O'Neill (2007).

Cultural as well as personal differences may also play a role in this connection and may be ethically significant. This becomes particularly important when society is composed of different cultural groups, because what is considered significant from one cultural perspective is not necessarily what is considered significant from another. Therefore part of the fiduciary obligation of the health care professional who interacts with a patient and tries to satisfy the latter's right to health care is to determine whether this would play a role in their decision making.

It is also important to note that it would be ethically inappropriate for a health care professional to assume that consent centres only in the outcome of a proposed intervention, and that the techniques or tools that might be used to try to achieve this outcome are of secondary importance. That may well be the central factor in the mind of the health care professional but, as will be discussed more fully later, it may not be what is central in the mind of the patient. Consequently it would be ethically inappropriate for a health care professional to ignore this possibility and simply to assume that the patient was only interested in the potential outcome of the proposed intervention when giving consent. Once more, then, this entails that the fiduciary relationship between professional and patient extends beyond merely health care related issues but includes the obligation of the professional to become aware of—and to respect—the value framework of the patient.

The process of obtaining informed consent also has important ethical implications. Thus, it is frequently assumed that if the patient or substitute decision maker signs a printed form on which the relevant information is outlined, then the patient or substitute has given consent. That, however, is not necessarily the case. While that may be acceptable in other areas—for instance, in commercial transactions—signing a form is insufficient for informed consent in health care. It is only if the patient or substitute actually understand, here and now, what is contained in the form that their signature is evidence that they have given consent. It may be difficult to achieve assurance in this regard. However, one way of doing so is to have the patient (or the substitute) repeat back, in their own words, what they understand about the proposed intervention in questions.

6.13 Types of Consent

Once again, however, there is more to the matter than appears at first glance. Consent itself comes in various forms. Thus, consent may be explicit or general, entailed, implied or assumed, and may even be current or anticipatory.

For instance, explicit consent occurs when the patient agrees to the intervention on the basis of understanding the nature of the specific intervention as well as its expected risks and benefits and the likelihood (or the probability) that the risks and benefit may or may not occur. General consent, on the other hand, is consent that is not for a specific intervention but for a type of intervention. For instance, unless it was specifically indicated by a patient what type of incision should be used, consenting to a caesarian section which is a general type of procedure would leave the precise nature of the incision that would be used undecided, and hence the consent to the caesarian section would be general in nature.

In both of these types of consent, the patient actually indicates something and is involved in the decision making process. In entailed consent, however, there is no specific patient decision about the actual intervention. Instead, consent to the intervention is entailed by something to which the patient has already consented. For instance, if a patient has consented to coronary bypass surgery, the patient is generally not asked to give consent to the fact that a general anesthetic will be used because such an intervention cannot be performed safely without a general anaesthetic. Likewise, consent to a phlebotomy in the sense of providing blood entails consent to a venipuncture, since engaging in a phlebotomy entails puncturing a vein.

Implied consent obtains when something that the patient does or says makes it clear that the patient understands the nature of the intervention but the patient does not specifically state their consent to the particular intervention in question. An example here would be a patient sitting down and holding out an arm for the phlebotomist or someone who is holding a needle in order to take blood from a vein—the presumption here of course being that the patients has already given informed consent for a blood test itself.

However, ethically, the notion of presumption itself has to be handled with care. For instance, the fact that a person agrees to be transferred to or to enter an emergency facility does not necessarily indicate they have consented to all of the types of interventions that may be considered medically appropriate and that would otherwise be delivered in that setting. The individual may simply want their pain, nausea or other specific problems alleviated but does not wish any life-saving interventions to be employed. This becomes particularly important in the case of geriatric patients who do not wish their lives to be extended but merely wish to receive medical procedures that would make for a "fair and easy passage" and who enter the medical facility on that basis. It is therefore ethically essential that the implication should not simply be in the mind of the health care professional.

To expand on this for a moment, it is often the case that healthcare professionals assume that because a patient is in a health care facility that they are asserting their right to health care and that they have consented to all the sorts of intervention that are usually provided in that setting. By and large, this is probably correct. However, it would be ethically inappropriate if the individual has on their person a directive that explicitly states that they would refuse a particular kind of intervention. An example of this would be the sort of case where the patient is brought into an emergency department after an accident or for some other reasons, and carries a signed document that states that under no circumstance would they agree to the administration of blood or blood products,[36] or that under no circumstances should pork or bovine derived medications be used.[37] The matter will be discussed more fully in a moment under the heading of "advance directives." For now, suffice it to say that even if such

[36] Porteri (2018).

[37] Blankinship (2018), Sattar et al. (2004).

a document was undated, it would still be ethically appropriate for a health care professional to accede to it simply because it would be the best indication of the patient's values that would be available at that point in time.

As to the distinction between a current and an anticipatory consent, consent is current if it is given at the time that the intervention is at issue and is valid as long as the situation under which the consent was given has not changed in any relevant detail. When the situation has changed in any relevant detail, the original consent would no longer be valid because the original consent was given on the basis of the patient's understanding of the situation as it existed at that time.

In this connection, it should be noted that the Principle of Autonomy also entails that except under very special circumstances, a competent patient has the right to withdraw their otherwise current consent because they have changed their mind. Therefore what was just referred as a change in circumstances includes this possibility as invalidating a previously given consent. The only sort of situation where this would not hold would be when the intervention that was initially consented to was in progress, and that stopping it would put the patient's welfare at greater risk than it was at the time the consent was originally given. The reason for this exception is grounded in the Principle of Nonmaleficence, which effectively stipulates that a health care professional has a duty to avoid becoming an agent of harm.

Anticipatory consent is also grounded in the Principle of Autonomy. As was said before, the Principle entails that, all other things being equal, a patient has the right to accept or reject any health care intervention. However, patients may be aware that they may not always be conscious or otherwise competent, and that therefore they may not be in a position to exercise that right. Therefore, as was indicated a few paragraphs ago, they may execute an "advance directive" which, as was also said, is really a consent that is given in anticipation of not being able to give an appropriate directive for what should then be done. Such advance directives are most commonly encountered in the case of individuals who have a particular religious or cultural background and who fear that they may be unable to indicate their wishes, but they are also increasingly common in the case of individuals who for some reason or other do not wish to be kept alive when they suffer from certain conditions but would be mentally incompetent to indicate this. In the latter case these directives are sometimes referred to as "living wills." Ethically speaking, such anticipatory consent is binding on the attending health care professional if it is specific to the type of issue at stake, and the professional would be ethically obligated to honour it unless the professional had reasonable grounds to suppose that the patient, when competent, had changed their mind.[38]

It is at this juncture, however, that the responsibility of the patient—which will be discussed more fully in Chap. 8—intersects with the responsibility of the health care professional. To put it in colloquial terms, patients have an obligation to "mean what they say." Therefore they have an obligation not to leave an advance directive in place if they have changed their mind. By the same token, health care professionals

[38] WMA Statement on Advance Directives ("Living Wills"); accessed 2020-10-27 at https://www.wma.net/policies-post/wma-statement-on-advance-directives-living-wills/.

are entitled to assume that if such a directive is still in place—either carried on the patient's person or available as part of the patient's health care record—it is still valid.

6.14 Substitute Decision Making

Finally, there is the question from whom the health care professional should receive consent.

That is to say, while the right to health care belongs to every member of society that construes health care as a right, not every member of society can meaningfully exercise that right either in general or on a given occasion. For instance, young children tend to lack the cognitive sophistication to fully understand what would be involved in a particular intervention even if this information were to be expressed in language that would otherwise be appropriate because the concepts themselves may transcend what can properly be expressed at that level. The concept of death may well fall into this category,[39] as may the concept of mental illness.[40] Moreover, young children tend to lack the emotional balance and the competently held values that are necessary to be able to make a valid decision. Similarly, patients who a mentally ill or who suffer from diminished intelligence may also not be able to exercise their right to health care on their own behalf. With due alteration of detail, similar considerations apply in the case of unconscious persons who do not have (or carry) an advance directive.

In these and similar cases, health care professionals are faced with a situation where a decision has to be made—i.e., a consent has to be given—lest the right of the patient to health care lapse because a consent is lacking. The question then becomes who should give that consent.

Traditionally, it was assumed that under such circumstances the fiduciary obligation of the health care professional entailed that the professional should make the decision on the patient's behalf. However, as careful analysis of the decision-making process in health care has shown, when a consent is ethically valid and is not merely an emotional reaction to a given situation, it is made on the basis of values. More specifically, giving consent is making a choice. However, merely being aware of options and understanding their implications is not yet to make a choice, as neither is reacting even if the reaction were to be consonant with the value system of the individual who reacts. Making a choice first requires understanding and then choosing among the available options on the basis of the values that one holds.

Values, then, may be described as concepts of the desirable with motivating force.[41] Consequently they function as action potentials and, absent values, choices

[39] Poltorak and Glazer (2006).

[40] Fox et al. (2008).

[41] Darwall (1983), Hyun (2002), O'Neill (2001), Hodgkinson (1996).

(which, as was said, are different from reactions) cannot and will not be made. Therefore the right to health care, which in turn incorporates the right to informed consent, is exercised on the basis of the values that are held by the right-holder or patient. Therefore unless health care professionals are aware of their patients' values and use them to ground the decisions that they would make on their patients' behalf, the decision that they would make would be based on the values that inform the decision-making of the professionals themselves. That—so it was realized—would not only be paternalistic, but would also be to hand over to the professionals the whole matter of consent. Consequently, it came to be realized that the decision should be made by a duly empowered substitute decision maker, and that—the Doctrine of Emergency excepted—this would not be the health care professional but someone who would be best placed to be aware of the values that the patient would use, and whom the patient could reasonably be assumed to accept as substitute decision maker.

The fiduciary nature of the professional-patient relationship, therefore, as it is currently understood, entails that the professionals should seek out this individual and transfer the consent process to them so that the individual's right to health care could be actualised. Ideally, of course, the professionals would have explored this issue with the patient while the patient was competent as part of the inception of their professional-patient relationship, but that may not always be possible. In particular, it may not be possible if the patient was never competent, as would be the case with young children or mentally disadvantaged persons or if the patient had previously declined to do so, or in the sort of emergency situation that was referred to above.

It is in the latter sort of case that the Doctrine of Emergency comes into play: If no duly empowered substitute decision maker is reasonably available and time is of the essence, then the Principles of Beneficence and Nonmaleficence entail that the health care professional should assume the role of substitute decision maker and actualise the patent's right to health care. However—and this follows from the intersection of the patient's right to give consent on the basis of their own values with the fiduciary obligation of the professional to respect those values—the values that the professional should then use should not be the values that the professional would use either as individual or on a professional basis, but the values that the objective reasonable person in the patient's position would use. This, of course, expands the realm of what is involved in the inception of a fiduciary health care professional-patient relationship and what is entailed by the obligation to act in the best interest of the patient: It means that it involves the obligation not only to take the patient's condition into account but also to become aware of the valuational framework in which the patient is situated—where, in the case of emergency contexts, this means becoming aware in a socially sensitive manner.

6.15 Records

A further series of issues that have ethical relevance for the right to health care and that is involved in how health care professionals are ethically obliged to fulfil their fiduciary obligations centre in patient records. This is becoming increasingly important in the contemporary setting, where the use of patient records has become an integral factor in health care professionals' lives. The issues include such matters as record keeping itself, their nature and communication, ownership, privacy and confidentiality, and related matters.

6.16 Record-Keeping Itself

As to record-keeping keeping itself, trivially it includes the duty to keep appropriate records for patient care.

At first glance, that seems to be a given. However, historically this has not always been understood. In fact, patient-specific records for the purpose of direct patient care are a relatively recent development. They only began to be systematically kept for this purpose in the latter part of the nineteenth century.[42] Before then, if patient-relative medical records were kept at all, they were kept mainly for accounting and peripherally for teaching purposes, and as such they were kept only by physicians.[43] It was only with the extension of health care professions in the nineteenth and twentieth centuries to include nursing, psychologists and technically focused health care providers, etc., that this came to include health care professionals in general. The advent of this development also went hand-in-hand with the increasing mobility of patients—mobility not only in the sense of moving from one geographic location to another but also in the sense of their changing from one individual health care professional to another—and the increasing specialization of health care services themselves.

6.17 Record Formatting

The latter in particular entailed that the data that were provided by the different health care specialties had to be incorporated into a single record that would provide a meaningful and accurate picture of the overall health status of the patient so that the patient could be treated in an appropriate fashion. This, in turn, entailed that the types of patient-relative records that health care professionals originated not only had to be accurate—or at least as accurate as possible under the circumstances that obtain—but also meaningful to the other health care professionals with whom they would be shared when it came to providing patient-relative health care services.

This became particularly important with the increasing expansion of patient care in terms of consults that are provided by specialized professionals, as well as with the advent of electronic health records and the introduction of eHealth and telemedicine.

---

[42] Kahn (1970), Reiser (1984).

[43] Siegler (2010), Franklin (1754).

While previously, with paper-based records it was possible to essentially ignore the need for standard formatting, this was no longer possible with electronic patient records. Electronic patient records (EPRs)[44] could literally not be shared unless the technical means of transmitting and receiving them were common to both sender and receiver. This meant that the technical transmitting format of the records had to be common to both sender and receiver.

Not only that: quite aside from the technical format of the records for transmitting purposes, the format in which the records themselves were set out and encoded had to be shared by sender and receiver. For, while an electronic handshake between sender and receiver might be possible if the technical transmitting format was the same, this did not guarantee that the format of the record itself as it was transmitted by the sender could be decoded by the receiver. If that was not the case, neither sender nor receiver would be able to meet their respective fiduciary obligations to wards their patients.

6.18 Patient Analogues and Mediated Professional-Patient Interaction

The advent of eHealth and telemedicine, in turn, raised a further and ethically much more fundamental issue and went much deeper than merely the use of EPRs: It went to the very ethics of the professional-patient relationship that is involved when health care professionals and patients relate to each other in this type of health care delivery.

That is to say, in traditional and so to speak standard sort of health care the interaction between patients and professionals is direct. By contrast, in eHealth and telemedicine the interaction between them is mediated and occurs through electronic means. As was already indicated, the question therefore becomes whether it is appropriate to assume that a fiduciary relationship arises between health care professionals and patients when such a mediated interaction occurs. The reason this is important is that if such a relationship can arise only on the basis of a direct interaction, then the professional would not be unethical if they did not respect a patient's right to health care in such a mediated interaction. In other words, if this sort of interaction did not give rise to the ethical aspects of a fiduciary relationship, then the services that the professional provided would fall under the rubric of commodities and of items for sale, and the considerations that were outlined when considering the health care professional-patient relationship above as being fiduciary in the ethical sense would not apply.

[44] In what follows, the term 'electronic patient record' is understood in the wider sense of electronic health record. While, strictly speaking, an electronic health record is more inclusive than an electronic patient record, the difference is irrelevant for present purposes. For more on this, see Kluge (2020).

It is therefore important to note that, as was pointed out in Chap. 5,[45] in fact a fiduciary relationship does arise even in such mediated interactions. To see why this is the case, it may be appropriate to restate what was said on that occasion and to begin by considering how such an interaction may take place. Thus, it may occur in various ways. For instance, the professional may interact with the patient by using an electronic tool such as a telephone or an on-line computer device; alternatively, the professional may interact with the patient by accessing the electronic record of the patient; as well, the interaction may take place both by using electronic communication tools and by the professional accessing the patient's electronic record.

In all of these sorts of cases however, as was also pointed out, when health care is considered a right a fiduciary professional-patient relationship is in fact established. In the first sort of case, the fact that the interaction between the professional and the patient is mediated through an electronic communication tool is in principle no different from the sort of situation when the professional talks directly with the patient or uses an x-ray machine or other device for examining the patient rather than doing so directly and physically by using their hands, ears, eyes or sense of smell.[46] The fact that the tool would be electronic in nature is irrelevant because that goes only to the physical nature of the tool, not to the fact of tool-use itself. After all, hands, eyes and related organs are also tools in the sense that they mediate data to the brain of the professional, even though they are organic in nature.

As to the second and third sort of case where interaction is mediated through the use of the patient's EPR, the fact that logically speaking the EPR is a patient analogue is sufficient to give rise to a fiduciary relationship.

The crucial term here, of course, is the term "analogue." While the notion of patient analogue has been discussed before, it may be worth repeating that an analogue is something that is similar or comparable to something else either in general or in some specific detail, and that therefore can logically function in place of that whose analogue it is.[47] An EPR is a patient analogue because it functions as *persona* of the individual whose analogue it is when the professional forms a diagnosis or assessment of the patient and makes a prognosis. In other words, it takes the place of the person itself in the interaction, and for that reason performs a similar function and is a substitute for the person of the patient. Therefore just as the direct and unmediated interaction between patient and health care professional gives rise to a fiduciary relationship, so the electronically mediated interaction between professional and patient gives rise to a fiduciary relationship. Interestingly enough, this is no different from what occurs when a primary care professional consults with another professional by providing the latter with the patient's record. In that sort of case, of course, there is a fiduciary relationship between the primary care professional and the patient. In a sense, the fiduciary relationship that the consult has towards the patient is derivative. However, the relationship is derivative only in a causal sense. Consequently the

---

45 Chapter 5, pp. 96 ff.

46 Bijland et al. (2013).

47 *Oxford English Dictionary*; *Meriam-Webster Dictionary*, etc.

combined approach that integrates the use of EPRs and telecommunication devices and techniques into the delivery of health care establishes a fiduciary relationship between health care professional and patient.

As an aside—and this takes the discussion back to the preceding chapter—it is important to note that in some cases it is ethically mandatory for a society to establish techniques of electronically mediated professional/patient interaction since only in this way can the health services that the society is ethically obligated to provide for its members be delivered. As the recent COVID-19 pandemic[48] illustrates only too well, access to appropriate health care, at least in its incipient stages, may well overwhelm the capacity of a health care system that relies on direct interpersonal interaction not only because of the sheer number of individuals who require health services but also because of the danger that this presents to the professionals whose duty it is to provide the relevant care.[49] Under such circumstances, a mediated access to health care may be the only way in which a society can meet its obligation to fulfil the right to health care of its members. With due alteration of detail, this even applies in normal times in the case of geographically distributed societies such as Canada, the People's Republic of China or Brazil where an insufficient number of health care professionals are situated in remote areas and direct interactions are not possible.

6.19 Privacy and Confidentiality

All of this—and in particular mediated health care—involves not only the use but also the sharing of patient records and this, in turn, raises the issues of privacy and confidentiality insofar as the health care professionals are concerned. More specifically, it raises the question of what privacy and confidentiality amount to for health care professionals when health care is construed on a socialized basis and when health care professionals function in such a setting.

It will be recalled from the previous chapter when the issues of privacy and confidentiality were examined from the perspective of society that the two notions are correlatives. Privacy in the informatic sense is a right that belongs to patients. It is the right—within limits—to keep their personal conditions, health status and relationships secret if they so wish, and to determine who shall have access to their records, to what parts of their records and under what circumstances, etc.[50] By contrast, confidentiality is a duty. It is a duty that belongs to whoever has the relevant patient information not disclose anything about the patients without the patients' consent (or, in the case of patient incompetence, without the permission of their duly empowered substitute decision maker), as well as the duty to protect the patient records in their possession against access by unauthorized parties.

[48] World Health Organization, Coronavirus disease (COVID-19) pandemic; accessed 20/11/2020 at https://www.who.int/emergencies/diseases/novel-coronavirus-2019.

[49] Webster (2020), Yada and Head (2019).

[50] For more on the limits to patients 'right of privacy, see Chap. 8.

It will also be recalled that the health care professional's duty of confidentiality has long been recognized,[51] and that in the traditional health care context this was relatively easily achieved. The health care professional could easily take appropriate material steps to ensure that no unauthorized party would have access to the patient records, and record encryption was really not an issue. However, in the modern context and with the use of EPRs, ensuring privacy requires putting in place appropriate and effective means to prevent accessing the place where the EPRs are stored—i.e., to prevent successful hacking—and to encrypt the records in case they might be successfully hacked.

Therefore a question that arises is whether the professional's fiduciary duty to protect patient privacy entails that the professional is obligated to personally ensure that the technical means for safeguarding the privacy of patient records are appropriate and effective or whether it belongs to some other party.

Arguably, the answer to this question follows the same logic as the answer to the question whether health care professionals have an ethical obligation to ensure that the consults that they obtain from other health care professionals are in fact valid and meet the otherwise appropriate standards. In each case, the obligation is to ensure that the expert to whom the task of providing a particular consult or of fulfilling a particular task is delegated is someone who is properly qualified. At the same time, however, the obligation of the professional to ensure privacy does not entail that the professional themselves has to acquire the relevant technical expertise or perform the relevant actions that would ensure that confidentiality would be maintained. Nonetheless it does mean that the professional has an obligation to take due care that appropriate and effective measures are in place, however that may be done.

6.20 Some Limits

However, as has already been pointed out in Chap. 5, both logically and ethically speaking such measures cannot be applied too far in an operational sense. In particular, no matter how health care professionals interact with their patients, when health care is provided on what has been called a rights-oriented or socialized basis they cannot provide appropriate health care services unless these services have been planned for at the policy level. That approach requires appropriate planning which sets out what health services society should make available to society's members and how to fund these services. Such planning, however, is dependent on data, and these data are ultimately contained in patient records.

Moreover, the delivery of health care involves the use of tools, medications and techniques. Without these, health care professionals also cannot carry out their mandate. However, as also was indicated in the preceding chapter, these tools, medications and techniques cannot be developed without scientifically valid research which again, directly or indirectly, depends on the data that are contained in patient

[51] Bruyn (1967), Edelstein (1943).

records. This applies not only to the records of participants in Phase 0 and I trials, which often do not involve actual patients, but especially to Phase III trials, where patients are necessarily involved.[52] Likewise, if health care professionals operate in an institutional setting, the institution has to plan for the services that the professionals will provide; and again, this is not possible unless the institution has access to the patient records.

Therefore, the sphere of privacy that otherwise surrounds patient records and with which health care professionals are obligated to surround their patients' records cannot be absolute but must be transparent to the degree that the relevant data in the records can be shared with policy makers, duly authorized researchers and by the institutions in which the professionals interact with their patients.

There are, however, certain limitations that apply both in the case of researchers as well as of policy makers with respect to the extent of their right to access patient records and the data that are contained in them. To be sure, both researchers and policy makers have data needs. However, there is a logical difference between the data that they need and the identity of the individuals from who the data are derived. Neither in the case of researchers nor in that of policy makers is it necessary for them to know the identity of the patients whose data they access in order for them to be able to carry out their respective tasks. Therefore while the health care professionals may share the records in question with these parties, the duty of confidentiality entails that they share them in anonymous form so that the identity of the patients will be protected. Therefore the standard approach that health care professionals are ethically obliged to take when they engage in such otherwise legitimate record sharing would not simply be to de-identify the data they contain but to anonymize them. The reason for that, in turn, is that simply de-identification may preserve informatic factors that allow the re-identification of the individuals from whom the records and the data that ae contained in them are derived,[53] whereas anonymization involves irreversibly severing the data from any items that might conceivably allow re-identification.[54]

6.21 Record Ownership

A further issue that arises for health care professionals in the context of records is that of ownership. It can be summed up in a simple question: Who owns the patient record?

Once again, the change in the health care professional-patient relationship from a paternalistic to a fiduciary perspective coupled with the advent of electronic record

---

[52] World Health Organization, *Clinical Trials*, accessed 25/11/2020 at https://www.who.int/health-topics/clinical-trials/#tab=tab_1

[53] Ribaric et al. (2016).

[54] Strictly speaking, this is not quite correct. As was pointed out above, it is possible to identify individuals on the basis of three data points even when data are anonymized and all personal identifiers are removed. However, the process is calculation intensive and requires appropriate computer resources. Cf. Porter (2008), El-Emam et al. (2009).

keeping and eHealth and telemedicine has changed the ethical understanding of the issue. Previously, and when health care was not construed as a right but as a commercial service, it was assumed that the professional who originated the record—and this applied particularly to physicians—owned the records as well as the data that were contained in them simply because the professional had originated the records. Ultimately, this position was based on the philosophical thesis that workers have an ownership-right to the product of their work.[55] However, the realization that health care was a right and that, ethically, the relationship between health care professional and patient is fiduciary in the sense that was outlined above, brought with it the realization that even though the professionals are paid for their services and even though it is part of their professional activity to generate both the records and the data that are contained in them, the records, being patient analogues, go with the patient.[56]

Ultimately, therefore, it was realized that the special status of patient records, no matter what their form, ethically entails that while the health care professional may claim ownership and control of the material instrument that contains the patient record as well as of their format—i.e., the way in which the data that are contained in it are arranged or set out—the record as informatic entity and the data themselves belong to the patient. They belong to the patient just as much the personal aspects that define the patient in real life belong to the patient. This has even come to be reflected in some legal provisions such as the European General Data Protecting Regulation.[57] With this, the issue of consent has come to extend not merely to the direct and indirect interventions on the patient as person but also to the use of their records, and the fiduciary obligations that health care professionals have towards their patients has come to include the duty to receive consent for the manipulation and use of their records.

Further, the fact that patients essentially own the informatic content of their records, together with the fact that the interaction between health care professionals and patients is on the basis of informed consent, means that patients (or their duly empowered substitute decision makers) have the right to see all aspects of their records. This holds true even if it means that patients might become aware of therapeutic errors on part of the professionals, because such errors should be noted in the records themselves since they are part of the overall picture of the interaction between the professionals and their patients. The only exception to patient access may be if the patient is incompetent. Then, however, the right of access transfers to the duly empowered substitute decision maker, who then would be exercising that right on the patient's behalf.

[55] Locke (1690).

[56] Kluge (1999).

[57] European Union, *General Data Protection Regulation*, accessed 04/11/2020 at https://gdpr-info.eu/.

6.22 Participating in Two-Stream Health Care

An issue that was touched on in Chap. 5 but that was not pursued any further was whether a health care professional's participation in a two-stream health care system would present ethical problems for that professional. It was then suggested that while the Principle of Autonomy applies to health care professionals as much as it does to any other person, there are certain reasons why it would probably be questionable that health care professionals could use such an argument in a rights-oriented health care system. However, the focus of that discussion then was from the perspective of society. It may therefore be appropriate to deal with the matter under the heading of health care professionals themselves.

The Principle of Autonomy is indeed to the effect that everyone has the right to self-determination. However, as has been pointed out on several occasions, it is not without conditions. More specifically, the right to self-determination that it underwrites is always "subject to the equal and competing rights of others." This means that autonomy is not license, and that the right to self-determination has to be seen in the context of the rights of others.

If the analysis in the preceding chapters is correct, then the rights of others include the right to health care. As was then pointed, out this right—which is underwritten by the Principle of Equality—entails that society has an obligation to provide health care services for health-based conditions that interfere with the ability of its members to take equitable advantage of the opportunities that are otherwise available in that society. However—so the argument continued—society can meet this obligation only if it provides its agents with the resources that are necessary for achieving this, and also only if it puts in place a public health care system to which everyone has access for their relevant conditions.

In and by itself this does not seem to affect the health care professionals' right to exercise their profession on a private basis. However, as was pointed out a moment ago, the right to self-determination is subject to the equal and competing rights of others. Therefore if functioning as a health care professional on a private basis interferes with the right to health care of members of society who cannot afford private health care—i.e., if functioning as a health care professional in a private fee-for-service health care sector interferes with the fulfilment of that right—then society, as supervenient authority, has the ethical right to imposes a limitation on the exercise of individual autonomy if acting in that fashion involves being a legally sanctioned licensed professional.

Of course, unless there was an ethical reason for society doing so, that would merely be a legal matter and not of ethical concern, and it would not have ethical implications for health care professionals. Moreover, the obligation to provide equitable access to health care belongs to society, not to health care professionals. Therefore unless a connection can be drawn between society's obligation on the one hand and the actions of health care professionals on the other, the claim that exercising their profession on a private basis would have ethical implications for health care professionals would remain unfounded. The question therefore becomes, whether there is such a connection.

And indeed there is. The fact is that health care professionals exercise their role within a social framework. From an ethical perspective, that framework is structured by a set of ethical considerations by which all members of society should abide. In other words, the ethical obligations that belong to a society form part of the overall social framework, and therefore they condition the domain of options that are ethically legitimate for its members. This is reflected in a society's provisions that control the interactions between its members where otherwise, in and by itself, it would be appropriate for them to exercise their autonomy in any way they wish. Therefore, if an individual's exercise of autonomy would interfere with society's ability to meet its ethically grounded obligations, then the individual's freedom of action may legitimately be curtailed within that social setting.

The right to health care in the sense that this has been discussed so far is not a right that belongs to persons in a vacuum, which is to say it does not belong to them simply because they are persons independently of their embedding. It is a right that arises in a social context. Its correlate is society's obligation to provide equitable access to health care for its members. It follows from what was said a moment ago that society therefore has the right to constrain the autonomy of its members—and that includes health care professionals—if their actions would interfere with its ability to fulfil this obligation. For reasons that have already been indicated in Chap. 5, a two-stream system of health care would seriously impair society's ability to do so. Since, as was pointed out above, an exercise of individual autonomy in a social context is ethical if and only if it is not at variance with the ethical framework that structures social interactions as such, it follows that it would be unethical for health care professionals to exercise their autonomy by providing health care on a private basis. In particular, it would be unethical if only duly licensed health care professionals may act in the capacity of health care providers.

However, if that condition was not met—i.e., if it did not impair society's ability to provide health care to its member son an equitable basis—then the conclusion would not follow. And, in fact, it seems that there are ways in which this consequence can be avoided. For instance, if the professionals provided health care services on a private basis only for those interventions that were not included in the domain of services that form part of the public health care system and, moreover, if they did so only in privately funded institutions. No conflict with society's equitable approach to health care would then seem to arise. Alternatively, if the professionals provided the whole spectrum of health care and did not use financial criteria when accepting patients but the same criteria of need as would be used as in the public sector, and if they used financial criteria only when providing health care that was not included in the public sector, this would also not seem to constitute an ethical conflict.

However, and aside from all else, there remains the fact that the knowledge and techniques that constitute part of a health care professional's armamentarium essentially are social goods.[58] That consideration would be strengthened even further if

[58] Cf. Chap. 5, p. 90.

part of the education of health care professionals was subsidized either directly by being provided to the individuals themselves or to the institutions in which they receive their training, as is the case in many jurisdictions.[59] It would also receive additional support if the setting in which they worked received support from society or some special accommodations. In any of these cases, it would impose an obligation on the professionals to accommodate their activities in a way that repays what in effect is a social debt.

Moreover—and this is probably the most telling factor—the central issue is whether practicing their profession in the private sector would interfere with society's ability to provide health care on an equitable basis. And here, as was suggested in the previous chapter, the fact that it would reduce society's ability to provide health care on an equitable basis because the reduced number of health care professionals that was available for engaging in the relevant services would be telling. Further, for reasons that were already mentioned, the fact that health care professions are service provider monopolies when they are licensed would exacerbate the situation. Consequently—possibly except for providing health services that would not fall under the equity condition that underwrites the right to health care—it is dubious that when health care is construed as a right, health care professionals are ethically entitled to practice their profession in the private sector. Of course, that would mean that if the professionals acted on this reasoning, if the the private sector existed at all it would be reduced to offering only very limited types of health care services, but that would not be an ethical issue.

6.23 Remuneration

A further issue that ethically limits the autonomy of health care professionals when health care is construed as a right is the matter of remuneration for their services. This was already discussed in the previous chapter, but there it was considered from the perspective of society itself. While the considerations that were then adduced find parallels in the context of health care professionals, it may be appropriate to restate them here in more specific terms.

To begin with, health care professionals in a non-socialized health care setting could decide at what level they should be remunerated and could attempt to enforce their position by threat of job action such as limiting or even withdrawing their services this. However, as was also pointed out in the preceding chapter albeit from the perspective of society, this is not the case when health care is considered a right and the relevant health care system is socialized in nature.

As to the size of the remuneration, once more the fact that society has limited resources here becomes an important factor. This is not to deny that, like all other individuals, health care professionals are entitled to receive due recompense for what they do, as well as for the time, training and preparing that goes into their

[59] Lee (1984).

being able to qualify for licensing and to act as health care professionals. It merely means that the size of the remuneration could not ethically be determined on the basis of purely economic considerations that centre in such things as scarcity of professional numbers, the perception on the part of members of society of what would be an appropriate level of recompense or related matters, etc. Nor would considerations that derive from comparison with the level of remuneration that other types of professionals receive be entirely appropriate unless these other types of professionals also practiced a monopolized profession that would draw for their recompense on a societal resource pool that was limited in extent, and where society itself had an obligation to provide the relevant services in an equitable manner to its members as a matter of.right. To proceed otherwise would be like comparing apples with oranges: The comparisons would be logically skewed.

As to the type of job action that health care professionals could ethically adopt in order to advance their claims this, as was suggested before, would be extremely limited because health care professions essentially are service provider monopolies. Therefore they could engage in limitation or even withdrawal of services only as long the health care needs of members of society would not be affected in any serious way. Ideally, of course, it merely means that both society and health care professionals would determine what remuneration would be appropriate by engaging in binding arbitration.

6.24 Conclusion

It should be clear from the discussion in this chapter that when health care is construed as a right in the sense that was outlined in Chap. 3, then the relationship between health care professionals and patients as well as between health care professionals and society itself acquires parameters that are importantly different from those that obtain when health care is construed as a commodity. However, construing health care as a right also conditions the rights of patients in their interactions with both health care professionals and with the society that underwrites this right. Moreover, the ethical relationship between health care institutions and society itself, as well as the nature of their relationship with both health care professionals and patients also assumes distinct aspects. The next chapter outlines the implications that the thesis that health care is a right has for health care institutions. The implications that it has for patients will be sketched in a subsequent chapter.

References

Aggarwal, A. 2010. The Evolving Relationship between Surgery and Medicine. *Virtual Mentor*. 12(2):119–23. https://journalofethics.ama-assn.org/article/evolving-relationship-between-surgery-and-medicine/2010-02. Accessed 15 Apr 2021.

Ancell, A., and W. Sinnott-Armstrong. 2017. How to allow conscientious objection in medicine while protecting patient rights. *Cambridge Quarterly of Healthcare Ethics* 26 (1): 120–131.

Bester, J., and E. Kodish. 2017. Children are not the property of their parents: The need for a clear statement of ethical obligations and boundaries. *The American Journal of Bioethics* 17 (11): 17–19.

Bijland LR, Bomers MK, Smulders YM. Smelling the diagnosis: a review on the use of scent in diagnosing disease. *Neth J Med*. 2013 Jul-Aug;71(6):300–307.

Bruyn, H.B. 1967. Confidentiality and the Use of Health Records. *Journal of School Health* 37 (4): 161–165.

Caldwell, J.A., J.L. Caldwell, L.A. Thompson, and H.R. Lieberman. 2019 Jan. Fatigue and its management in the workplace. *Neuroscience and Biobehavioral Reviews* 96: 272–289.

Carrez, L., L. Bouchoud, S. Fleury, C. Combescure, L. Falaschi, F. Sadeghipour, and P. Bonnabry. 2019 Sep. Work overload is related to increased risk of error during chemotherapy preparation. *Journal of Oncology Pharmacy Practice* 25 (6): 1456–1466.

Carrin, G., and C. James. 2005. Social health insurance: Key factors affecting the transition towards universal coverage. *International Social Security Review.* 58 (1): 45–64.

Childress, J.I., and M. Siegler. 1984. Metaphors and Models of Doctor-Patient Relationships: Their Implication for Autonomy. *Theoretical Medicine* 5: 17–30.

Combs, M.O., R.M. Antiel, J.C. Tilburt, P.S. Mueller, and F.A. Curlin. 2011. Conscientious refusals to refer: Findings from a national physician survey. *Journal of Medical Ethics* 37 (7): 397–401.

Cowley, C. 2017. Conscientious objection in healthcare and the duty to refer. *Journal of Medical Ethics* 43 (4): 207–212. https://doi.org/10.1136/medethics-2016-103928.

Darwall, S. 1983. *Impartial Reason.* Cornell University Press.

Edelstein, Ludwig. 1943. *The Hippocratic Oath: Text, Translation and Interpretation*, 56.

El-Emam, K., F.K. Dankar, R. Vaillancourt, and RoffeyT and Lysyk M. 2009. Evaluating the Risk of Re-identification of Patients from Hospital Prescription Records. *The Canadian Journal of Hospital Pharmacy* 62 (4): 307–319.

Emanuel, E.J. 1991. *The Ends of Human Life: Medical Ethics in a Liberal Polity.* Cambridge: Harvard University Press.

Fox, C., E. Buchanan-Barrow, and M. Barrett. 2008. Children's understanding of mental illness: An exploratory study. *Child: Care, Health and Development* 34 (1): 10–18.

Franklin, B. (1754). *Some account of the Pennsylvania Hospital: From its first rise to the beginning.* Philadelphia: Franklin & Hall; available at https://collections.nlm.nih.gov/bookviewer?PID=nlm:nlmuid-2554043R-bk

Freedman, B. 1975. A moral theory of informed consent. *The Hastings Center Report* 5 (4): 32–39.

Goethe, E.R. 1989. Lawyers as Officers of the Court. *Vanderbilt Law Review* 42 (1): 39–91.

Hyun I. 2002. Waiver of informed consent, cultural sensitivity, and the problem of unjust families and traditions. *Hastings Center Report* 32 (5): 14–22.

Hodgkinson, C. 1996. *Administrative Philosophy.* Pergamon, *passim.*

Kahn, A.P. 1970. From Ben Franklin's vision to medical record reflections. *Medical Record News*: 42–55.

Kleisiaris, C.F., C. Sfakianakis, and I.V. Papathanasiou. 2014. Health care practices in ancient Greece: The Hippocratic ideal. *Journal of Medical Ethics and History of Medicine* 7: 6.

Kluge, E.-H. 1999. Medical narratives and patient analogs: The ethical implications of electronic patient records. *Methods of Information in Medicine* 38 (4–5): 253–259.

Kluge, E.-H. 2020. *The Electronic Health Record: Ethical Considerations.* Elsevier: Academic Press.

Lee, R.H. 1984. Subsidizing the Affluent: The Case of Medical Education. *Journal of Policy Analysis and Management* 3 (2): 276–284.

Locke, J. 1690. *Second Treatise of Government*, II, Sect. 6.

Manson, N.C., and O. O'Neill. 2007. *Rethinking Informed Consent in Bioethics.* Cambridge U Press.

Marketos, S.G., and E. Poulakou-Rebelakou. 1995. Tradycyjna medycyna w starozytnej Grecji (współistnienie sztuki asklepiejskiej i medycyny hipokratejskiej) [Traditional medicine in ancient Greece (coexistence of Asclepian art and Hippocratic medicine)]. *Przegl Lek* 52 (12): 612–4. Polish. PMID: 8,834,663.

Montgomery, J. 1988. Children as property? *The Modern Law Review* 51:3, 323–342.

Mousourakis, G. 2007. *A Legal History of Rome*, Routledge, pp. 33 *et pass.*

Nock, A.D. 1952. The roman army and the Roman religious year. *Harvard Theological Review* 45: 187–252.

O'Neill, O. 2001. Practical Principles and Practical Judgment. *Hastings Center Report* 31 (4): 15–23.

Pfaff, H. 2004. Surgical safety and overwork. *British Journal of Surgery* 91 (12): 1533–1535.

Poltorak, D.Y., and J.P. Glazer. 2006. The development of children's understanding of death: Cognitive and psychodynamic considerations. *Child and Adolescent Psychiatric Clinics of North America* 15 (3): 567–573.

Porter, C.C. 2008. De-identified data and third party data mining: The risk of re-identification of personal information. *Shidler J. l. Com. & TecH.* 5: 3.

Porteri, C. 2018. Advance directives as a tool to respect patients' values and preferences: Discussion on the case of Alzheimer's disease. *BMC Medical Ethics* 19 (1): 9.

Blankinship, L.A. 2018. Culturally Sensitive Care, for Islamic Patients and Families. *Journal of Christian Nursing* 35 (2): 94–99.

Reiser, S.J. 1984. Creating form out of mass: the development of the medical record. In *Transformation and Tradition in the Sciences: Essays in Honor of I. Bernard Cohen*, ed. E. Mendelsohn, 303–16. Cambridge University Press.

Ribaric, S., A. Ariyaeeinia, and N. Pavesic. 2016. De-identification for privacy protection in multimedia content: A survey. *Signal Processing: Image Communication.* 47: 131–151.

Savulescu, J. 2006. Conscientious objection in medicine. *BMJ* (332): 294–297.

Sattar, S.P., et al. 2004. Patient and physician attitudes to using medications with religiously forbidden ingredients. *Annals of Pharmacotherapy* 38 (11): 1830–1835.

Siegler, E.L. 2010. The evolving medical record. *Annals of Internal Medicine* 153 (10): 671–677.

Sim, J., and R.B. Purtilo. 1991. An ethical analysis of physical therapists' duty to treat persons who have AIDS: Homosexual patients as a test case. *Physical Therapy* 71 (9): 650–655.

Tang, C., C. Liu, P. Fang, Y. Xiang, and R. Min. 2019. Work-related accumulated fatigue among doctors in tertiary hospitals: A cross-sectional survey in six provinces of China. *International Journal of Environmental Research and Public Health* 16 (17): 3049.

Tauber, A.I. 2005. *Patient Autonomy and the Ethics of Responsibility.* Cambridge: MIT Press.

Veatch, R.M. 1972. Models for ethical medicine in a revolutionary age. *Hastings Center Report* 2: 5–7.

Webster, P. 2020. Virtual health care in the era of COVID-19. *The Lancet* 395: 10, 231, 1180–118.

Yada, N., and M. Head. 2019. Attitudes toward health care virtual communities of practice: survey among health care workers. *Journal of Medical Internet Research* 21 (12): e15176.

Yakubu, A., M.O. Folayan, N. Sani-Gwarzo, P. Nguku, K. Peterson, and B. Brown. 2016 Apr. The Ebola outbreak in Western Africa: Ethical obligations for care. *Journal of Medical Ethics* 42 (4): 209–210.

Chapter 7
Health Care Institutions

Abstract The purpose of the present chapter is to consider how understanding health care as a right affects the conduct of institutions that actualize that right both at the organizational and the operational level. It further considers such institutional issues as privacy, research and consent protocols, as well the ethical issues that are presented by the fact that different types of professionals are involved in their operations.

Keywords Allocation · Consent · Research · Operational issues · Organization issues · Professional issues

7.1 Introduction

The preceding chapter dealt with some of the ethical issues that face health care professionals when health care is construed as a right. The purpose of the present chapter is to consider how understanding health care as a right affects—or more correctly perhaps, how such an understanding *should* affect—the conduct of institutions that actualize that right at the establishment level, where "institutions" is understood in the sense of organizations that are formally organized into administrative structures that subscribe to certain principles and provide the administrative framework within which hands-on health care is delivered.

The actions of health care institutions may consist in providing health care in a more or less localized setting, as for instance in a hospital that provides health care services in a given catchment area, or they may consist in providing health care services that transcend such geographic limitations, for instance when the catchment area that they service is geographically extended. That notwithstanding, the reason health care institutions deserve separate consideration is that no matter whether health care is construed as a right or as a commodity, it is they who provide the structural framework within which the care is actually delivered when the scale of interaction extends beyond the limited social setting of single professionals interacting with individual patients, as was discussed in the previous chapter.

In some ways, of course, the issues that arise for these health care institutions overlap with those that have already been discussed in Chap. 5 with respect to society as a whole, and in Chap. 6 with respect to health care professionals. While health

E.-H. W. Kluge, *The Right to Health Care: Ethical Considerations*, The International Library of Bioethics 92, https://doi.org/10.1007/978-3-030-93838-3_7

care institutions are like societies in the sense that they are administrative structures and share some of the problems that face all administrative structures, the issues that they face are not at the higher-order policy level that is characteristic of society's involvement; and while just like societies they do have an obligation to provide health care, that obligation is derivative and much more limited in scope than is the case with societies themselves. Moreover, while they employ the health care professionals who actually engage in the relevant actions at the hands-on level and the ethical issues that they face overlap with those that are faced by these professionals—consent is here a telling example, as is the issue of privacy—these issues differ in some important ways from the way in which they arise for the professionals.

The focus of the present chapter, therefore, is to sketch and consider some of these issues in greater detail.

7.2 General Considerations

To set the stage for this, it may be appropriate to begin by reiterating certain facts. Perhaps the most important of these is that in a society that consists of more than a few individuals, a health care system cannot fulfil its function—indeed, cannot exist as a system—unless it provides the implicated services at the hands-on level in an organized manner and in a context that integrates the relevant kinds of health care services into more or less unified settings in which the professionals who deliver the services can interact with their patients. This is particularly the case if the services are specialised in nature, require specialised equipment or involve distinct professional specialisations. Therefore the existence of institutionalized health care structures is as essential for a society to meet its health care obligations as is the presence of individual health professionals.

These institutionalized structures, in turn, face challenges both at the organizational and the operational level. At the organizational level, one of the more profound challenges is that they must integrate the actions of different kinds of professionals so that their respective frameworks do not conflict with each other; at the operational level, one of the more challenging issues is that while they must deal with questions that face all institutional structure, they must do so in a manner that is consistent with the overall ethical requirements that frame a rights-oriented health care system.

7.3 The Organizational Level

More specifically, one of the organizational challenges that an institutionalized deliverer of health care faces is that the delivery of that care involves more than simply the health care professionals who provide patient care at the hands-on level. It also includes other professionals whose expertise lies not in health care but in the areas of technical and bureaucratic services. In other words, it also includes administrators and managers as well as technicians and other support workers who enable the institution to function. The problem is that these other parties are not bound by the same considerations as those that bind health care professionals. And when the health care institution is embedded in a society that sees health care as a right, integrating these parameters in the context of a rights-oriented framework presents this organisational challenge in a way that differs from how it arises when health care is construed as a commodity.

As though this was not enough, as soon as an institution's venue extends beyond the localized setting, issues arise that in many ways differ from the way that they arise for health care institutions that function solely at a local level. And when an institution draws on services from different jurisdictions—as is sometimes the case when health consults are involved and local experts are unavailable—the problematic becomes even more complicated, because different statutory requirements may be involved and different ethical considerations may be deemed relevant in the respective jurisdictions. This is particularly the case when the institution is embedded in a rights-oriented jurisdiction but the other jurisdiction does not subscribe to that perspective, since the original institution still has to satisfy the ethical requirements that ground the right to health care itself.

Moreover, as was pointed out in Chap. 6, when health care is construed as a right, health care professionals have an ethical obligation to always act in the best interest of their patients insofar as these are consistent with the overall rights-oriented framework, whereas when it is construed as a commodity their obligation is inherently dictated by purely professional interpretation of fiduciary obligations and by the nature of the contractual arrangement that they have with their patients. However, no matter whether health care is considered to be a right or a commodity, whatever actions the professionals can perform is functionally dependent on the tools that are available to them. For instance, when dealing with a patient who suffers from cardiac insufficiency different actions would be professionally mandated when surgical interventions are available as opposed to when they are not. Likewise, diagnosing a particular condition like breast cancer when sophisticated cellular and electronically based technology is unavailable forces the professional to rely on hands-on techniques like palpitation or to refer that patient to another setting; and all of this has implications for what would count as acting in the patient's best interest. All of this has to be taken into account by the institutional framework in which they operate.

Moreover, what counts as satisfying the best-interest criterion that governs health care professionals is also determined by reference to professional consensus which, among other things, takes into account the current state of the art relative to the instruments or tools that are available. In the present context, this means that a publicly funded health care institution has an obligation to adjust, in accord with these consensus-based standards, what tools and instruments it makes available to the professionals whom it employs so that they can in fact act in the best interest of their patients and in keeping with these standards. Therefore, institutionally speaking, a constant feedback mechanism is ethically mandated to ensure a harmonious integration of tool-availability and standards. As well, given the evolving state of health care technology, a publicly embedded health care institution ethically has an obligation to ensure that appropriate measures are in place so that so-called "legacy" tools and instruments—tools and instruments that are outmoded—are identified and replaced as and when appropriate.

By contrast, when health care is construed as a commodity, both what counts as being in the patient's best interest and what tools or techniques should be employed in the institutional setting is essentially a matter of contract negotiation and litigation. While the latter may make reference to consensus-based standards, these would not be ethically binding. As was said, the institutional obligations in this regard—insofar as they arise at all—would not be a matter of ethics but of contract, negotiation and litigation.

7.4 Valuational Complications

A further organizational issue that faces health care institutions that function in a rights-oriented setting, and in particular an issue that is likely to arise in a multicultural or multi-religious society, is whether—and if so, how—they should accommodate in their organizational structure the valuational perspectives that exist among the members of the society whose health care needs they are to supposed to serve, since these differences may well affect the kinds of health care services that would be considered acceptable.

As an aside, this would not be an ethical issue for a health care institution that was embedded in a health care system that treats health care services as commodities. Thus, such an institution could decide to adopt a religiously based pro-life value framework and only offer services that would be in keeping with its own religious tenets, or it could adopt a culturally-based health care perspective and only offer those services that would be in line with its particular cultural tradition; and it could do so even though the services that it would not offer would fall under the rubric of health care services to which members of the same society would be entitled on a rights-based perspective. Of course any such institution might thereby run the risk of alienating those members of society who did not share these religiously or culturally grounded value perspectives. However, seeing that the institution would be situated in an overall framework that treated health care services as commodities, this would not present an ethical problem. The institution's position would here be no different from that of any other corporation that offered a particular product or service but that limited the range of its products or services on the basis of its value considerations.

However, a health care institution that was embedded in a rights-oriented system could not adopt such a standpoint because it would be obligated to provide otherwise appropriate health care services solely on the basis of need, as this was outlined in Chaps. 3 and 4, and to align its organizational framework in keeping with a particular set of religious or cultural values would be to deprive those who did not share those values of the services to which they would be entitled as a matter of right. Consequently, for a health care institution in a rights-based health care system to adopt such an approach would be for it to violate the underlying ethics of the rights-oriented health care framework itself. As to those members of society who subscribed to values that would find such services objectionable, they could simply decide not to access them. However, their values would not determine the range of services that was offered by the health care institution.

7.5 Specialization of Services

In this connection, it is important to note that this issue is different from the question whether a rights-oriented health care institution—and this applies in particular to health care institutions that are intended to provide health care in a particular catchment area—may ethically decide to specialize and not provide the whole gamut of health care services to which the population would otherwise be entitled.

The analyses in the preceding chapters suggested that it would indeed be ethically acceptable as long as doing so was defensible in terms of the Principles of Impossibility and Nonmaleficence. That is to say, trivially, what services a health care institution can offer—where this includes both material and professional parameters—is functionally dependent on the resources that are available to it. Therefore it may be practically impossible for a particular health care institution to defray the cost of providing the tools, devices or facilities that are necessary for all types of health care services without compromising the latter's quality. Likewise, it may be impossible for it to offer a particular type of health care simply because the professionals who would be needed to engage in the relevant actions would be unavailable in that particular location even if appropriate transportation or consultation methods were to be employed. An example of this would be the case with cardiac transplantation when the specialists who are professionally qualified to engage in that intervention are situated in a different location. While the institution would have the obligation to arrange for the transfer of the patients who needed such an intervention to an institution where such a service are provided, the decision not to provide that particular kind of care and to specialize only in the kinds of services that it could provide in a qualitatively acceptable manner would be ethically defensible because doing so would be based on the Principles of Impossibility and Nonmaleficence.

7.6 The Operational Level

At the operational level, however, slightly different ethical issues arise. To begin with, a rights-oriented health care institution encounters challenges that centre in parameters that affect its existence and functioning in a material sense.

In some ways, of course, these challenges are universal and are faced by any institution simply because they are parameters that affect any institutional structure. For instance, efficiency and effectiveness are operational issues that arise in any institutionalized context and that have to be addressed in any type of institution because any type of institutional structure only has a limited amount of resources.

However—and unlike institutions that are not engaged in providing health care—an institution that provides health care in a rights-oriented setting may not deal with these issues by considering practicality, efficiency and effectiveness as the primary determinants of its operations, whereas these would be central for institutions that do not function in rights-oriented setting. Of course, while practicality, efficiency and effectiveness of operation are also relevant considerations that a rights-oriented health care institution has to take into account, it is the ethical considerations that are entailed by the construal of health care as a right that would ultimately be decisive in its operational decisions and may, on occasion, have to carry the day. So, for instance, while such an institution would be obligated to maximize the efficiency of its operations because it could not pass any efficiency-related costs to its patients—something that explains why health care institutions of the public type tend to be operationally more efficient than private health care institutions[1]—it cannot ethically do so in any way that impairs or detracts from its offering of the designated services on an equity-oriented basis.

Continuing for the moment in the same vein, a health care institution that functions in the public sector is financially dependent on society for its funding and cannot charge for its services the way a private health care institution can. Instead, it has to rely on the funds that it receives from the society that establishes it. This means that it cannot ethically defray any cost overruns that it might experience in its operations by charging its patients for any additional health-related costs—something that may be possible for a private health care institution, which can frame its contracts accordingly. It must either reduce the types of services that it offers or their number, or both. This, however, while it is ethically defensible if it is done for the reasons that were outlined above, is something that may run into problems with respect to the mandate that it has been given by society.

7.7 A Modified Two-Stream Approach and Its Implications

An obvious question to ask at this point is whether a health care institution that operates in a right-to-health-care framework may offer, as elective and paid-for options, interventions that do not fall under the rubric of socially guaranteed rights simply in order to augment its financial resources. Such interventions might include things like non-essential surgery whose purpose would be to satisfy a personally held beauty standard, or treatments whose purpose would not be to ameliorate health-based conditions that would interfere with an individual's ability the take advantage in an equitable manner of the opportunities that society offers to its members but to increase the individual's ability to compete successfully in professional sports events.

[1] Woolhandler et al. (2004).

The obvious reason for considering this question is, of course, that branching out in this manner might well be financially advantageous for the institution because it could then charge for such services. It might even charge its patients for upgraded non-medical services such as gourmet meals, entertainment or an improved general ambience. As was said, when properly done, this might enlarge its pool of resources so that it could expand the array of health care services that are socially mandated and which theoretically it is supposed to provide but that it cannot provide because of financial constraints.

Arguably, for a health care institution to adopt this option would be to operate in what in Chap. 5 was identified as a two-stream approach to health care. However, while adopting a two-stream approach at a societal level would run into the problems that were raised in that discussion, it is important to remember that a health care institution is not society itself, and that the problematic that it faces and the considerations that are relevant to its operation are not quite the same as those that face society. For instance, an institution does not have to address the issue of professional availability in quite the same way as society because, unlike society, it does not have a duty to ensure that there are a sufficient number of professionals to allow for equitable access to health care so that all relevant interventions could be staffed. Arguably, therefore, it seems that as long as the institution was able to satisfy the health care needs of the patients who fall within its own domain of operation, and as long as providing the extra services would not interfere with that ability, there would seem to be no ethical objection to adopting this mode of operation.

At the same time, and appearances notwithstanding, this reasoning may not be quite as telling as at first glance it may seem, because it would ignore the implications of the underlying assumption that gave rise to these considerations: more specifically, the assumption that the institution exists in a rights-oriented framework. As such, the institution's operational structure has to be in accord with the principles and provisions that condition the ethical framework that applies to the rights-oriented society as whole. In other words, it could not consistently choose an option that would interfere with the conditions that are laid down by, and that shape the nature of, that framework.

This then plays out as follows: As was pointed out in Chap. 5—and indeed as it obvious on the slightest reflection—the number of qualified health care professionals in any social context is limited. Therefore to offer this limited number of health care professionals the option of working in what would essentially be a two-stream setting would be to contribute to a reduction in the overall number of health care professionals that would be available for society to meet its obligation of providing equitable access to health care for its members. Consequently, for an institution that was embedded in a rights-oriented health care framework to structure its setting into a two-stream approach for health care professionals would be for it to adopt a position that stood in direct contradiction to the ethics of the rights-oriented approach in which it is embedded.

Further, the issue of the availability of professionals themselves, whether these would be directly involved in providing hands-on health care or in providing support services, would not be confined to having a sufficient number of qualified professionals for the institution's functioning in terms of numbers. It would also extend to the issue of their time.

That is to say, professionals—as indeed any person—can reasonably be expected to work for only a certain number of hours, and when these are exceeded not only does the risk of adverse effects on the professionals themselves increase, but the risk of the professionals making errors in their treatment of patients and hence the risk to the welfare of the patients increases as well.[2] Both of these present problems for a health care institution that followed a two-stream model, and would have to be addressed. It is unclear how this matter could be resolved without curtailing the availability of health services in general and thus negatively affecting society's ability to fulfil its right-to-health-care obligation.

Moreover, the issue of time—i.e., of temporal availability—would extend beyond merely the availability of professionals and support staff. It would also extend to the use of equipment and space. More specifically, operating in a two-stream manner would require that the institutional facilities in which the professionals' activities would be carried out when they would be engaged in the private stream, as well as the tools that they would be using, would be unavailable for providing care in the public-sector stream. Therefore, unless this was taken into account when the institution itself was established so that a sufficient number of rooms and equipment was made available without interfering with the institution's ability to take care of its public-stream patients, and unless both the maintenance and the operating costs that would arise in this regard were somehow funded in a way that did not draw on the financial resources of the institution *qua* public health care institution, it would once more impact in a negative way on the institution's ability to fulfil its public sector mandate. In the end, therefore, it seems highly questionable that a two-stream approach within in a public sector health institution is ethically defensible.

7.8 Informed Consent

Another important issue that arises for a health care institution in operational terms is the issue of informed consent. As was pointed out in Chap. 6, all other things being equal—as was then also explained[3] and as will discussed further in the next chapter,[4] exceptions to this are limited—obtaining informed consent for any intervention is ethically mandatory for health care professionals before proceeding with any action, and therefore would necessarily be included in the protocols of an institution that employs them. This requirement, so it was argued, is grounded in the Principle of Autonomy, which is to the effect that everyone has the right to self-determination subject only to the equal and competing rights of others. This is not only a matter

[2] Gottlieb (2003), Wen et al. (2016), Hills (2015).

[3] Chapter 6, pp. 120 f.

[4] Chapter 8, pp. 172 ff.

of ethics but also integral to global policy provisions like the *Universal Declaration of Human Rights* and is clearly and explicitly enunciated in Article 5 of that *Declaration*[5]:

> The autonomy of persons to make decisions, while taking responsibility for those decisions and respecting the autonomy of others, is to be respected. For persons who are not capable of exercising autonomy, special measures are to be taken to protect their rights and interests.

More precisely still, the *Declaration* then stipulates in Article 6 of that.

> Any preventive, diagnostic and therapeutic medical intervention is only to be carried out with the prior, free and informed consent of the person concerned, based on adequate information. The consent should, where appropriate, be express and may be withdrawn by the person concerned at any time and for any reason without disadvantage or prejudice.

With due alteration of detail, and along the same lines, similar provisions are found in most international pronouncements and national provisions that deal with consent to health care. This, therefore, suggests that protocols for informed consent should to be integral to the functioning of all health care institutions and certainly for health care institutions that are embedded in a rights-oriented health care system.

However, while this is correct ethically speaking, it may be problematic in an operational sense for such health care institutions. The reason is that the notion of person varies from socio-cultural context to socio-cultural context. For instance, as was discussed in Chap. 2, ethics and law currently tend to frame the notion of a human person in terms of an individual who is distinct from all others. This is the sort of conception that is reflected in the European Union's notion of "natural person" around which it structured its *General Data Protection Regulation*,[6] and it is central to the Supreme Court of India's finding that the right of privacy is a fundamental human right that belongs to each person.[7]

While on this conceptualization a human person may be embedded in a social context, such an individual is a distinct and unique entity and their embedding affects neither their identity nor their autonomy. Therefore on this understanding a person is free to involve a group such as their family in their decision making, but whether they do so is a matter of their choice. As "natural persons" they are free to make their own decisions and to choose their own values, and to use these values in their own decision making. On this understanding, therefore, the issue of informed consent would be satisfied by an institution structuring its consent protocols in such a way that a material and distinct individual would be presented with the relevant information and then made a decision, and the protocols could be structured without reference to

[5] UNESCO, *Universal Declaration on Bioethics and Human Rights*; available at http://portal.unesco.org/en/ev.php-URL_ID=31058&URL_DO=DO_TOPIC&URL_SECTION=201.html.

[6] European Union, *Regulation (EU) 2016/679 of the European Parliament and of the Council on the protection of natural persons with regard to the processing of personal data and on the free movement of such data, and repealing Directive 95/46/EC* (General Data Protection Regulation) https://eur-lex.europa.eu/eli/reg/2016/679/oj.

[7] Supreme Court of India, *Writ Petition (Civil) NO. 247 OF 2017*; accessed 23/02/2021 at https://uidai.gov.in/images/news/Supreme_Courts_Order_in_WP_247_277_304_of_201716062017.pdf.

the fact that the individual might be embedded in a particular social grouping such as a family or an extended family.

However, as was indicated a moment ago, there are alternative conceptions of personhood according to which human persons are not what one might call atomic or solitary entities who are distinct and separate from all others. Rather, they are conceived of as entities who by their very nature are intimately interwoven into the particular social fabric or collectivity in which they are embedded. In other words, they are conceived of as what might be called embedded entities.[8] Therefore on this perspective, autonomy is not simply centred in the isolated individual but encompasses the collectivity in which the individual is embedded; and not only is their own decision-making guided by the values of that collectivity, but decision making itself is a shared affair. Therefore the conception of autonomy that is integral to the approach that was outlined in the preceding paragraphs would find little resonance in socio-cultural communities that had this embedded understanding of personhood; and to insist that autonomous decision making would be satisfied if the steps that were undertaken relative to the biological and prototypically solitary entity on the basis of the atomic understanding of personhood would be to invalidate the legitimacy of such a distinct socio-cultural perspective. Differently, on this embedded construal of personhood the Principle of Autonomy and the matter of informed consent would not be satisfied by institutional consent protocols that only followed the atomic conception of personhood.

The question, therefore, that would face a health care institution in a rights-oriented social context that served a multicultural population is whether—and if so, how—this conception of embedded personhood should be accommodated in its consent protocols.

Fortunately, the solution to this conundrum would seem to be provided by the Principle of Autonomy itself. Procedurally, it would lie in adjusting the conditions that govern a patient's initial entry into the institutional framework. More precisely, and from a procedural perspective, it could commence upon entry of the biologically distinct individual for whom decision-making becomes relevant—and that would essentially include all patients—into the institutional framework. It is at this juncture that it could be determined to which model of personhood the individual subscribed; and it is on this basis that distinct perceptions of autonomy and their divergent approaches to consent (and privacy) could then be accommodated. Likewise, if an individual who entered the health care institution lacked capacity, it could be determined at this juncture to which model the individual could reasonably be expected to subscribe. It is at this point, then, that the applicable decision making protocols would diverge into two branches. One would be geared to the solitary or atomic model, the other to the embedded model. Correspondingly, privacy protocols would diverge and would be attuned to the ethical concerns of the respective models.

Arguably, one of the shortcomings of this approach would be that it assumes that there would be sufficient time to determine to which model of personhood the individual subscribes or could reasonably be expected to subscribe. That is to say,

---

[8] Hardwig (1990), MacIntyre (1981), Sandel (1982).

it may be objected that while what has just been outlined might be workable when entry into the institutional setting is on a non-emergency basis, it would not work in emergency contexts because there would be no time. Moreover, if the individual in question lacked capacity, as is frequently the case in such situations, someone who might provide the relevant information may be unavailable.

However, this difficulty is more apparent than real. To see why, it may be useful to recall that a similar problem exists in the context of religious perspectives that prohibit certain types of medical interventions for their adherents. Jehovah's Witness, Christian Science adherents, members of the Judaic or Islamic faith, etc. are here implicated. In such cases, the generally accepted solution is for society to stipulate that if the adherents of such faiths wish to be exempted from standardly accepted health care interventions or to be diverted from generally accepts modes of treatment, then they should carry with them a legally recognized advance directive which specifies that such should be the case. A similar approach is adopted in jurisdictions that insist on opt-out provisions for organ donation.[9]

In other words, the ethically appropriate factor that would provide a solution would be that of an advance directive. This device has even found legal acceptance in most modern jurisdictions. When the logic of this device is applied to the matter at issue, it has the advantage of acknowledging the overarching sovereignty of the jurisdiction in question while at the same time not invalidating the ethics of the personhood-perspective of the distinctive socio-cultural or religious groupings. It would therefore be consistent with the ethical principles and provisions that are enshrined in the *Universal Declaration of Human Rights*.

7.9 Privacy

Another operational issue that an institutionalized health care structure has to deal with is the issue of privacy.

Privacy was already identified as an ethical issue in Chaps. 5 and 6. However, in Chap. 5 it was discussed from a societal perspective and relative to the data needs that arise in the context of developing appropriate policies for actualizing the right to health care, and relative to the issues that arise when the delivery of health care extends beyond the localized setting and involves so-called mediated interactions between healthcare professionals and patients. In Chap. 6 it was discussed from the perspective of what implications the health care professional's duty of confidentiality has with respect to records that are under their control or in their possession. In the present context, whose focus is the institutionalized setting, essentially the same issues arise but they arise in a somewhat different way and assume distinct parameters in their own right.

[9] Cohen (1992).

7.9.1 Preliminary Considerations

A useful way to approach the issue of privacy in this context is to begin with the concept of data needs.

To begin, then, with the notion of data needs itself: As was said before, and purely logically speaking, the institutional setting in which health care is delivered is an administrative structure, and as such has data needs. In this it is no different from the administrative structure that characterizes a health care system as this was considered in Chap. 5 from a societal perspective. However, it differs fundamentally from the latter in that unlike the societal framework, it is an administrative structure that is actually and directly engaged in the delivery of health care. This means that its data needs are much more specific both in an operational and an organizational sense than those of society. Society's data needs arise at a much higher level because they do not deal with specific individuals suffering from specific conditions but with types of individuals, types of conditions, and with the formulation of policy.

In keeping with this, it might be argued that health care institutions do not need to concern themselves with data like the incidence and prevalence of health care conditions within society as a whole in order to fulfil their roles in a rights-oriented health care system. These are data that society needs in order to be able to structure an overall health care system that meets its obligation to provide its members with equitable access to health care services in an equitable fashion, but they are not data that a health care institution needs in order to fulfil its mandate as an institutional health care provider.

However, to argue thus would be to ignore the fact that a health care institution in a rights-oriented framework has a catchment area of persons whose health care needs it is supposed to meet. Moreover, it would be to ignore the fact that it could not fulfil its mandate if it focused only on the health conditions as they exist at a given moment in time because, given the increased mobility of patients, planning to meet the health care needs of the population in its catchment area would also have to take this mobility into account. Likewise, it would have to take into account changing patterns in the health profiles of the individuals in the catchment area that it serves, as well as changes in the nature of the health conditions themselves as these are presented by its catchment population. Emerging disease patterns are of particular concern in this regard. Moreover, it would have to factor the availability and qualifications of health care professionals who could function in its particular setting into to the design of its service delivery structure. Therefore an ethically conscious health care institution in a rights-oriented framework would have to cast its data net almost as widely as that of society itself, even though in certain respect it would be more specific.

Moreover, all of these factors have financial implications. After all, providing any of the relevant health care services involves expenditures. Again, purely logically speaking this is not much different from what faces society. However, once again, the institutional financial data needs are much more specific than those of society. They involve details of the financial implications of providing particular amounts and types of health care services, as well as the everyday expenses that are involved in actually running the institution in all of its material aspects. Moreover, they include the expenses that are incurred by employing the professionals and the support staff who are engaged in actually providing the services that it offers.

7.9.2 *Data Needs*

As was said, like any other kind of institution, a health care institution is not a haphazard affair. Quite aside from requiring the resources that are necessary for it to exist and function, it requires information both about those who actually provide the services that it offers as well as about the individuals—the patients—whom it serves.

7.9.2.1 Professionals and Employees

Beginning with the issue of information about the professionals that are involved in the actual delivery of health care, the data needs that a health care institution has with respect to the health care professionals whom it employs or who have privileges to work in its setting assumes different aspects from those that are encountered by society. Thus, a health care institution not only has to have information about the qualifications of the professionals whom it employs but also about their availability. While qualification standards in general fall under the auspices of society itself, a health care institution has to have information that is specific to the qualifications of the particular professionals who actually deliver the health care services in its setting, as well as information about their individual availability to deliver the relevant services in a timely and appropriate manner.

Further, because the institutional setting in which health care is actually delivered is a material setting, this involves more than only health care professionals and more than only data about them. It also involves material upkeep considerations and expenses of the sort that encountered by all other institutions. Further, because it is a health care institution, it also has to take into account hygienic and related considerations. This means that the need for non-medical staffing and their qualifications to engage in hygienic and related matters also has to be considered. Even information about the qualifications and the availability of technical staff who specialize in the maintenance and operation of the electronic and other tools that are used by the health care professionals has to be available to the institutional decision-makers if the institution is going to be able to function.

Furthermore, contrary to what is the case with society, all such information has to be identified information. It goes almost without saying that the institution has an obligation to protect all such data against any access or communication without explicit permission of the subjects of the data—an obligation that is grounded in the Principle of Autonomy. Therefore a health care institution's duty to protect the privacy of data about the individuals in its employ is just as strong as the duty that it has to protect the privacy of data about its patients.

7.9.2.2 Patient Data

This brings the discussion to the matter of patient-relative data and their privacy and security protection.

Directly or indirectly, the person-relative data that a health institution legitimately requires and has in its possession for its functioning are ultimately grounded in its delivery of health care. Thus, in one way or another, it is the patients' conditions and treatment that give rise to the various administrative and operational factors that have been mentioned. Trivially, therefore, it follows that the administrative structure that is the institutional health care setting cannot function unless it has access to patient data.

Patient data, however, come in various forms. There are identified data—i.e., data that are specific to a particular patient and carry an explicit patient identifier and which, simply by becoming known, would directly allow the identification of that patient. Then there are deidentified data—i.e., data where the identifier that would allow the direct identification of the patient has been deleted but where, when appropriately analysed, the re-identification of the patient would still be possible. Pseudonymized patient data also fall into this category. And then there are anonymized data—i.e., data where any specifics that indicate about whom they are have been deleted and where it is in principle impossible to re-identify the individual in question.[10]

Identified data, in turn, may contain different types of identifiers. They may include direct identifiers, such as names, addresses, phone numbers, social security numbers, medical record numbers, etc.; or they may contain indirect identifiers such as date of birth, gender, zip codes and similar features. Then there are pseudonyms and encrypted identifiers, which themselves do not identify the individual directly but that can be used to determine the identity of the individual from whom the data were derived by linking the data themselves with data from other lists.[11]

At this juncture, and as an aside, it may be appropriate to note that apparently even anonymized data are not necessarily absolutely immune from re-identification. Some authorities have argued that given access to three distinct data about an individual that are anonymized, re-identification is possible if the data are linked to prescription records combined with some medical data.[12] It has even been suggested that if anonymized patient health data are related to sufficiently large and varied non-health care related data sets such as sets that involve business transactions, internet searches, smart devices, employee lists, industrial equipment, videos, social media,

[10] Ribaric et al. (2016).

[11] European Union, *General Data Protection Regulation*; available at https://gdpr-info.eu/

[12] Emam et al. (2009), Porter (2008), Richardson et al. (2015).

voter lists, etc., their re-identification is possible.[13] If that is correct, then the question therefore becomes whether privacy in the strict sense can really be guaranteed by a health care institution unless techniques of quantum cryptography are used to ensure anonymization. And even then, it may not be possible to guarantee privacy unless the relevant data are contained in sets that are physically distinct and cannot be connected. For, while quantum encryption and computing can guarantee that privacy will not be breached undetected, it cannot guarantee that it will not be breached.

Returning, then, to the matter of institutional obligations, what all of this means is that when protecting patient privacy, health care institutions have an ethical obligation to take these parameters into account and to structure the privacy protection of the patient records in their possession accordingly. This, in turn, means that they would not only be obligated employ appropriate and effective measures that apply to the informatic structure of the records themselves—encryption has already been mentioned—but also institute appropriate and effective access measures that govern who may legitimately access the records and under what conditions. It goes almost without saying that the measures that they adopt to ensure this would have to be made known to the professionals who are involved with the handling of patient records.

7.9.2.3 Control and Possession

Before going any further, however, it may be useful to say a few words about record control and record possession, because these concepts are relevant to the ethical functioning of health care institutions in a rights-oriented system in an operational sense.

Patient records are considered to be in an institution's possession when they are contained in the informatic devices that are part of the institution's material infrastructure; they are considered to be under its control when they are either contained in its material infrastructure or in devices such as servers that are distinct from the institution's material infrastructure but that only the institution itself can legitimately access.

As was pointed out before, health care institutions cannot function unless they have the data that are necessary for their operations. Consequently, by the Principle of Impossibility, when a patient enters a health care institution the institution itself has the right to establish appropriate records about that patient. However, given that patient records essentially are patient analogues—a matter that was already touched on in Chap. 6 and that will be discussed further in Chap. 8—it follows that patient records are surrounded by a sphere of privacy which, all other things being equal, may be breached only with the consent of the patient or of their duly empowered

13 Na N, Yang c, Lo C, et al. Feasibility of Reidentifying Individuals in Large National Physical Activity Data Sets From Which Protected Health Information Has Been Removed With Use of Machine Learning; accessed 07/01/2021 at https://jamanetwork.com/journals/jamanetworkopen/fullarticle/2719130; Rivers et al. (2013), Sobhani and Saxon (2019) HIPAA Isn't Enough: All Our Data is Health Data. USC Center for Body Computing, accessed 07/01/2021 at https://medium.com/@usccbc/all-our-data-is-health-data-57d3cf0f336d.

substitute decision maker; and that it may be breached non-consensually only if it is demonstrably necessary to protect the health of others or, alternatively, if it is demonstrably necessary in order for the institution itself to be able to function and deliver the services that are its mandate. This, then, means that a health care institution has an obligation to ensure in its informatic protocols that the patient records that are in its possession or under its control can be accessed only by duly empowered persons and only for legitimate reasons—which, as was said, means that it has an obligation to structure its informatic framework accordingly.

A general rule that applies in this regard is that all and only those individuals who play a functional role in a health care institution and whose role requires that they be able to access these records should be able to do so. Therefore an ethical institution has a duty to put in place appropriate and effective safeguards which, all other things being equal—the qualifier will be explained in a moment—would prevent anyone other than such individuals from being able to access them. These safeguards have a dual nature: one is procedural, the other is structural.

7.9.3 Procedural Considerations

The focus of the procedural safeguards is the identity of the individual who seeks to access the patient records and the role that they play in the functioning of the institution.

More specifically, each individual who is employed by a health care institution performs a specific task. That task, in turn, carries with it certain specific types of data needs. This is obvious relative to health care staff who are directly involved with patient care because they have to adjust what they do in accordance with the condition and needs of the patients. However, it also applies to individuals such as janitors because they have to know whether the patients whose environment they clean are infectious, are psychiatrically challenged, etc., as well as to housekeepers, dieticians and to others, so that they can adjust their actions accordingly.

In purely logical terms, therefore, each such type of activity is associated with a particular domain of data that satisfies the data needs of those who are engaged in that type of activity. This may be called their characteristic data-need domain. These data-need domains may overlap, as for instance the data-need domain of a cardiac surgeon will overlap with that of an anaesthesiologist, or that of a psychiatrist with that of a hospitalist, and all of these may overlap with that of the clinical nurses who take care of patients. Nevertheless, they will not be exactly coextensive, and each will have a central data need domain that is unique to the specialization in question.

Accordingly, the procedural safeguards that a health care institution is ethically obliged to put in place should allow automatic access to the data that meet the typical data-needs of the type of professional who seeks that access. They would then block access if the data fall outside of that domain. However, they should not block such access absolutely unless the data in question lie entirely outside of the data-need domain that is usually associated with the profession of the individual who seeks the access. So, or instance, an attempt by a custodian or an accountant to seek access to the hematological status of a patient would fall entirely outside of their professionally grounded data-need domain, and hence should be blocked absolutely. Likewise, the attempt by a cardiac surgeon or a nurse to access the record of a patient who was not in their care should also fall under the same restriction. On the other hand, if the data are the data of a patient who is under their care but fall outside of the data-need domain that is usually associated with their profession, then access should be allowed but should be flagged and brought to the attention of an appropriately positioned administrative security authority that was established by the intuition, so that the individual who attempted to do so would have to justify their actions.

The point is worth emphasizing. Ethically speaking, the privacy protection that a health care institution is ethically obligated to put in place should not be such as would be structurally absolute except in the sorts of cases that were just indicated. In the first instance, the data need domains that are attached to a given type of action are domains whose extent has been identified on the basis of past experience and according to the generally accepted criteria of those who engage in that type of activity. However, a professional may have good reasons to suppose that an innovative approach for dealing with a particular situation may produce better results than the usual kind of approach, and that it may be more effective. That, however, might require access to data that are not included in the data-need domain that is usually associated with the type of activity.

Moreover, emergencies and unforeseen situations may arise in the health care setting, and dealing with them may require a similar extended and unusual access by the professionals if they are to meet their fiduciary obligation towards their patients. Consequently the procedural safeguards that a health care institution establishes to protect the privacy of patient records should include an emergency override function in order to deal with these types of situations. Still, even in such cases it would be ethically obligatory for the institution to make sure that any such emergency override would not only be flagged and recorded but would also have to be justified to the appropriately placed institutional authority. Moreover, for reasons that will be explained more fully in the next chapter, the patient whose record had thus been accessed would also have to be informed of such an occurrence.

7.9.4 Structural Considerations

Nor is this the end of the matter. Health care institutions also have an obligation to structure how the patient data that are under their control or in their possession are kept. The reason for this is that this has ethical implications; and if the mandate of the institutions is to satisfy the patients' right to health care, this should not be at the expense of the right to privacy that belongs to the patients.

For instance, patient data may be stored in devices that are owned by the institution and that are located on the premises where the institution provides patient-oriented services. Alternatively, however, they may be stored in devices that are owned by the institution but that are not located on their premises and that have to be accessed using telecommunication technology. Still differently, they may be stored on servers that do not belong to the institution—in other words, they may be stored "in the cloud"[14]—and that can be accessed only through the use of telecommunication technology. And so on. While each of these methods of patient data storage has to satisfy the privacy considerations that were raised above under the heading of procedural issues, they also present distinct and ethically important structural issues.

One of these—and independently of where the records are stored—is simply the fact that electronic patient records are stored in electronic devices. This makes them susceptible to privacy breaches through the very act of accessing itself even when that access is itself legitimate. This is not true of material records. While the privacy of material patient records can be more or less readily safeguarded by putting in place material safeguards that prevent access by unauthorized individuals,[15] this is much more difficult when they are kept on a server no matter where the server is located or who owns or controls it. The reason for this is that a server has to be accessed using a key-board or some such device by means of which a connection is established with the server on which the record is stored. However, the keystrokes themselves that are used to access the server and the record are subject to surveillance both electronically and visually. This is colloquially known as keystroke logging.[16] Consequently, health care institutions have an obligation to put in place appropriate measures to prevent such things as unauthorized keystroke logging. Failure to do so would not only run the risk of compromising the privacy of patient records but would also run the risk of allowing the unauthorized modification or even erasure of such records. If the latter were to happen, the institution and the professionals who work in that setting would no longer be in a position to meet the patient's right to health care, thereby invalidating the very reason for and justification of their existence and professional operation.

As to accessibility, patient records that are not accessible as and when they are needed by duly authorized professionals who are engaged in looking after the heath care needs of patients might just as well not exist. Trivially, therefore, the institutions in which the professionals interact with their patients have an obligation to ensure that the relevant records are available and accessible when they are required, since otherwise the patients' right to health care could not be met.

This obligation, in turn, has distinct but complementary aspects. The first is that the devices that are necessary to access the records should be kept in good functioning order, and that the means for accessing them—i.e., the relevant electronic communication devices—should be appropriately placed and functionally available as and when they are needed. The second aspect is that the format in which the

[14] "In the cloud" refers to the use of servers, networks, storage, development tools, etc. through the internet.

[15] United Nations, *Records and Information management guidance*. "Protecting records in offices and storage areas"; available at https://archives.un.org/sites/archives.un.org/files/RM-Guidelines/guidance_secure_records.pdf.

[16] Kaspersky. "What is Keystroke Logging and Keyloggers?" available at https://www.kaspersky.com/resource-center/definitions/keylogger; Nyang et al. (2014), Sagiroglu and Canbek (2009).

records are stored should be kept up-to-date and consistent with the platform that is used by the health care professionals when attempting to access the records from the material confines of the institutional setting.

However, the fact that a patient is physically located in a particular health care institution and that the health care professionals who provide the relevant health care services are affiliated with the institution does not mean that the professionals will always interact with the patient or query the patient's record when the professionals are actually located in that institutional setting. The professionals may wish to—indeed, for various reasons they may have to—do so when not located on the institutional premises. This, however, is possible only if the electronic format in which the patient records are kept by the institution and the electronic platform of the records are compatible with what is used by the professionals when they try to access the record from an external location. Among other things, it also means that the records should not be kept in what is called a legacy format. A health care institution, therefore, has a corresponding obligation to arrange its data protocols so that the relevant formats and platforms will mesh.

A final factor—and this once more falls under the rubric of availability—is that patient records must exist in order to be available. Trivially, if they do not exist they cannot be accessed, which in turn means that it would be impossible to provide the health care to which patients are otherwise entitled and which the institution is obligated to provide. Consequently there lies an institutional obligation to guard against the unauthorized erasure or destruction of patient records.

This, however, is not simply confined to providing appropriate back-up measures and facilities in case of the material malfunctioning of the devices on which the records are kept, or for situations where disasters in one form or another destroy the storage devices or facilities themselves and the records with them. The non-existence of patient records may also take the form of ransomware that encrypts electronic patient records and thereby effectively makes them non-existent as far as the institution and its professionals are concerned.[17] Consequently, an ethical health care institution is obligated to put in place appropriate means to prevent such an occurrence.

7.10 Tools, Devices and Facilities

Records, however, are not the only technical or material matters relative to which health care institutions have ethical obligations when they are part of a health care system that recognizes a fundamental right to health care.

In a rather obvious sense, a patient's right to appropriate health care cannot be fulfilled in an institutional setting if the tools, devices and facilities that are instrumentally necessary for doing so are unavailable to the professionals who work in that institution, or when they are not qualitatively up to par. The reason for stressing

[17] Alert (AA20-302A) (2020).

this is that when health care is construed as a commodity, the issues of tools, devices and facilities is a matter of contract, and any deviations from what is considered "normal" and even from what is standardly expected is a matter of negotiation and, if it comes to that, a matter of litigation.

Returning, then, to the case of health care institutions that are situated in a rights-oriented health care system, the obligation to provide appropriate tools, devices and facilities does not mean that such health care institutions are obligated to provide these for all types of services that are implicated by the general right to health care. As was indicated before, not every such institution is obligated provide such things as surgical cardiac interventions, chemotherapy or neonatal intensive care. Even though the institutions may be parts of an overall system of health care that construes health care as a right, they may specialize and address only certain types of health care needs, and only provide the tools, devices and facilities for dealing with the latter.

However, no matter what the area in which they may specialize and no matter what their catchment area that they serve may be, they do have an obligation to make certain that the tools, devices and facilities that are necessary for providing the type of health care that they claim to provide are not only sufficient in nature but also qualitatively appropriate and are available when the occasion demands. Moreover, they have an obligation take appropriate steps to ensure that the public in their catchment area is aware of what will and what will not be available in that particular setting.

The last comment may benefit from some further clarification. Health care institutions, even when they consist of several buildings, are generally situated in a particular location. As such, they serve the health care needs of the individuals who are located in their particular catchment area or who enter that catchment area and whose health care need arises on an emergency basis. Consequently appropriate planning for such occasions is an integral task of institutional management. At the same time, since a given health care institution may not be able to provide all the services that the individuals in their catchment area may need and to which they are otherwise entitled as part of their socially guaranteed right to health care, the institution has an obligation to make sure that the fact and the precise nature of the limitation in the services that that they offer is publicly known and, moreover, that appropriate measures are in place to the transfer individuals whose health care needs would not be met by the institution but that fall under the rubric of socially guaranteed right to health care to another institution where they will be met.

As well—and this returns the discussion to the matter of institutional obligation to provide appropriate tools and devices for health care professionals—publicly situated health care institutions have an obligation to provide the health care professionals who are part of their service structure with the tools, devices and facilities that are considered appropriate by duly constituted bodies of the relevant professions, and that they are kept up-to-date and in good working condition. Moreover, just as in the case of electronic patient records, they have an obligation to replace legacy devices on a continual basis and as the situation warrants, and to update and keep in functioning order the facilities that the professionals require. With due alteration of detail, this also applies to the medications and similar substances that are used in their setting since these may also be considered "tools of the trade".[18]

7.11 Research: From Society to Institutions

A health care institution cannot address the health care needs of its patients successfully if the modalities that would be necessary for doing so do not exist. Moreover, even when the relevant modalities do exist and even when they are available, a given intervention may not be effective to the degree that would constitute equitable access to health care because its level of success falls below what is the case with other interventions. If that is the case then, as was suggested in Chap. 3, while on those occasions society has an obligation to still provide appropriate care, it also has a concomitant duty to engage in research in order to rectify the situation.

There are several ways in which society may discharge that obligation. For instance, it may establish and fund agencies or organizations that function under its jurisdiction and purpose them to engage in research in the relevant areas. Alternatively, it may purchase the research from privately owned agencies or corporations and then integrate their results of into the design of its operational framework. Still differently, it may involve the health care institutions that are already funded by it and which, so to speak, are its agents—in other words, the health care institutions that have been the subject of the preceding discussion—and enjoin them to conduct the relevant research as part of their mandate.

When all is said and done, however, it is arguable that the last-mentioned option is ethically the most acceptable. There are several reasons for suggesting this. Thus, the analysis in Chap. 3 notwithstanding, the first approach would require separate and distinct funding, which in turn would diminish the resources that would otherwise be available for delivering health care. A similar argument could be raised against adopting the second option, with the additional consideration that it would be to inject a private and for-profit stream into what supposedly is a rights-oriented approach to health care. Only the third option would essentially avoid these difficulties for a rights-oriented approach to health care because—and this expands on the reason for rejecting the second option—any such research ultimately has to involve patients.

To clarify the ethical difference between these options, it may be useful to outline what is involved in this sort of situation a bit further. When health care is private in nature and health services are commodities, access to patients for the sake of conducting health related research seems to be essentially a matter of contractual arrangement, and patient participation in such research would seem to follow the same lines as involvement in any venture in the private sector. However, even in

[18] Additionally, they should be similarly concerned with the qualifications of their staff, and have appropriate protocols to guarantee that the individuals who carry out their profession on their premises or under their jurisdiction meet the relevant, appropriate and current professional standards in their actions.

context of private health care, sooner or later patients would be involved, and it is this fact that assumes distinct significance. For, there arises the question whether the patients should be remunerated for their involvement in the research as this would be the case with participants in any other research project in the private sector—and if so, whether the remuneration should be confined merely to the hope of being involved in something that would address their own health care issues or whether it should involve financial considerations that would be attuned to the severity of the relevant condition. Further, there arises the question whether the participants—i.e., the patients—would or should have a stake in the financial returns that the corporations would reap if the research project was successful. Likewise, the issue of ownership could arise, given that in some cases—particularly in cases where genomic properties of the patient-participants are involved and the product itself incorporates some of the genomic characteristics of the patient-participants. The case of breast cancer research leading up to the development of the BRCA1/2 tests is here a case in point.[19]

Many of these and related issue would not play a role when health related research is conducted in a health care institution that was situated in a rights-oriented setting. The ultimate aim of such research would not be the financial rewards that could be reaped if the research turned out to be successful but the improvement in the health care services that the institution could deliver to its patients. As to the matter of patient reward for participating in the research, it would not even enter the picture; and as will be discussed more fully in the next chapter when considering the issue from the perspective of patients, participation in research could be seen as a kind of duty that falls on patients as a condition of their being able to exercise their right to health care in any real sense.

Focusing, then on research in a rights-oriented health care institution, the question becomes how precisely such an institution would structure its research protocols.

7.12 Research Phases

For the sake of conceptual clarity and to lay the foundation for the analysis that follows, it may be useful to briefly recall what is involved in research in health care before dealing with the matter itself,.

It is generally accepted that ethically appropriate process standardly involves: Phase I–III trials.[20] While Phase 0 trials are the preliminary stages to determine whether the physiodynamics of a pharmaceutical hold out promise of effectiveness, and while they are mandatory in some jurisdictions, it is with Phase I trials that research properly so-called actually begins. Phase I trials are trials in which the safety and potential side effects of the given product or intervention are evaluated on healthy volunteers. Phases II and III involve actual patients. Therefore even though

---

[19] Ghosh (2012).

[20] World Health Organization, *Clinical Trials*; accessed 08/10/2020 at https://www.who.int/health-topics/clinical-trials/#tab=tab_1.

all of these phases involve human beings as subjects, it is really Phases II and III that involve patients who actually suffer from the conditions for whom a successful or improved method of treatment is sought; and while in general the hope is that the participants in these Phases will derive a therapeutic benefit from the experimental interventions, the direct intent of involving them is not to provide therapy but to test the effectiveness and appropriateness of the intervention in question.[21]

7.13 Effectiveness Considerations: A Recap

As society's agent, health care institutions have a derivative obligation to provide their patients with health care modalities whose purpose is to return or to keep them in a health state that allows them to participate in an equitable manner in the opportunities that the society offers its members. While offering the best available care may go some way towards doing so, if doing so is not effective then it cannot be counted as fulfilling that obligation.

To expand on this for a moment—and this draws on some of the considerations that were raised in the previous discussion of success and effectiveness—unless there are logical reasons to suppose otherwise, the Principle of Equality entails that the effectiveness or the success-rate of the health care interventions that fall under the rubric of socially guaranteed health care interventions should be the same for a given population. If they differ, then, the Principle entails that appropriate steps should be taken to adjust them accordingly: Not, however, by selecting the population to which such interventions are offered—in other words, not by engaging in creative statistical manipulation—or by lowering the effectiveness ratio of the other interventions, but by increasing the effectiveness of the interventional outlier.

However, as was said, the health care modalities that fall under the rubric of socially guaranteed health services and that an institution may provide are not necessarily equally effective in their application. There may be several reasons for this, ranging from co-morbidities from which the patients suffer to age and sexual differences, and may even include genetic differences between distinct patient sub-populations. Indeed, the latter in particular is the reason for the recent drive towards personalized medicine.[22]

Still, it is possible to identify a given set of interventions that are considered to be effective in their application on the basis of current standard reviews and of professional consensus and which, minimally, satisfy the equity-criterion that underlies the social obligation to provide health care in the first place. With respect to these interventions, it is possible to identify their range of effectiveness for the society's population. Using this as the standard of what counts as a successful or effective intervention, it is possible to apply this value range to the interventions that are provided and thus to identify those that are not considered to be successful because

[21] For a different opinion, see Miller and Brody (2003).

[22] Di Sanzo (2017), Chatelin et al. (2017).

they fall outside of that range. These, of course, will be the interventions for which remedial interventions or remedial alternatives are initially employed. However, in their case the equity condition that underlies the obligation to provide health care in the first place entails that suitable steps should be taken to remedy the situation. The only way to do that, however, would be to engage in research. Consequently, the Principle of Equality entails that in all such cases health care research would become mandatory for a society that accepts the Principle of Equality—to say nothing of the Principle of Beneficence.

7.14 The Institutional Role

This returns the question to the issue of what role, if any, rights-oriented health care institutions have in all of this.

To put it succinctly, their role is not only to alert society of the fact that research is needed but also to provide the venues where such research can take place. It is at this juncture, then, that the societal and institutional frameworks one once more intersect: most particularly with respect to funding, for the initial funding that society provides to health care institutions to enable them to provide socially mandated health care would not include funding for research. This would be considered a separate funding item that society accords the institutions that are engaged in such research.

As to providing the venue for research, it derives from the fact that the participants in that research would still be patients, and therefore not only have the need for but also the right to appropriate care. Furthermore, the institutional setting is well-suited to the controlled environment that is necessary for conducting scientifically valid research.

7.15 The Institutional Protocol

However, as with any health care intervention, informed consent is a necessary condition for involving patients in research. This assumes special importance when the patients find themselves in an institutionalized setting, because they may think the delivery of the care that they would otherwise receive in that setting may depend on their agreeing to becoming involved in the research. An ethically structured research protocol of an institution, therefore, would minimize and if possible eliminate the likelihood of such a perception.

An ethically appropriate way for a health care institution to do this ideally would be not to include the professionals who treat the patients (and who stand in a fiduciary relationship to them) in the research protocol. Certainly, however, it would be to arrange the consent process in such a way that it was neither initiated nor conducted by the professionals who would take care of the patients but by someone else, and that the latter was unlikely to be perceived by the patients as being in a position of power over them. Moreover, it should come in two stages. In in its initial stages it should take the form asking the patients whether they would agree to let their name stand as possibly being included among the names of those would agree to participate in the relevant research. In other words, the institutional protocols for this initial consent—if indeed consent was given—should clearly state that it was not consent to being included in the research itself, but merely to allow their names (or identifiers) to be considered for inclusion in the research population. It should make clear that a separate and distinct consent process would be required if their name was selected from the list of potential research subjects.

As to the second stage of consent—i.e., consent for being actually included in the research—the institutional consent protocol should state clearly, and in language that the prospective participants understand, that if there was a intervention that was standardly applied to deal with their condition, they would either receive that standard intervention as so-called active placebo or the intervention that was being researched; and that if no standard intervention existed for their condition, a harmless intervention that would not have any detrimental effect on their current condition—i.e., a pure placebo—would be used in place of the otherwise standard intervention. Further, participants in the research would be told that which treatment the individual participants would receive would be determined by choosing among the names of the patients who had agreed to act as participants, but that this would be done on a random basis so that neither the researchers nor the participants would know which individual received what kind of intervention. In other words, the protocol would clearly indicate that the actual research would be randomly selected double-blind and. placebo-controlled in nature.

How the resulting data would be handled and who would have access to them would also be part of the consent process and the protocol, and the institution would take the same steps be to ensure the privacy of any patient-relative data. As well, it would clearly indicate that any use of the data beyond what was agreed to in the initial consent would be subject to a distinct consent.

Finally, for reasons that are grounded in the Principles of Equality and Autonomy, the institution would distinguish between cognitively compromised patients and those who could give consent on their own behalf.

That is to say—and as will be explained further in the next chapter—because cognitively compromised patients are just as much persons as anyone else, they have an equal right to health care. Therefore, despite their being cognitively compromised, their inability to give a valid consent should not exclude them from possibly being participation in research: all the more so since the fact of being thus compromised may itself be a factor in the disease process or the condition that is being researched. Their possible participation, therefore, should be determined by their duly empowered substitute decision maker—as indeed is the case in any matter of consent to possible treatment, or even relative to their admission to the institution itself. However, it would be part of the institutional research protocol that it would be necessary to seek the agreement of such compromised individuals insofar as they would be able to give agreement. In other words, as grounded in the Principle of Autonomy, it would be necessary to structure the institutionally approved protocols in such a way that, if at all possible, the compromised populations would have to give what is sometimes called assent. Moreover, if they gave any indication that they would no longer wish to be included in the protocol once the research had started, that this would be sufficient to terminate their participation in the research.

7.16 Conclusion

The preceding discussion has attempted to highlight some of ethical issues that health care institutions face when dealing with patients, professionals and with society itself if they are situated in a rights-oriented social setting. As was indicted at the beginning of the discussion, even though these issues overlap with those that face society and professionals as these were discussed in the preceding chapters, they arise in different ways and have different implications. However, as should be clear from the tenor of the present discussion itself, such differences do not call for ethically distinct solutions but for solutions that are consistent with the fundamental Principles that govern the ethical frame work of a society that recognizes the fundamental right to health care.

Not all issues that face health care institutions have been adequately addressed, however. For instance, the issue of substitute decision making for compromised patients that enter an institutional setting has not really been considered, as neither has the issue of exceptions to the informed consent requirement or the question of the ethical implications of privacy relative to the genetic data which play such a crucial role in the research and in the delivery of personalized medicine. Issues such as this, however, are more profitably addressed from the perspective patients. This, then, will be the focus of the next chapter.

References

Alert (AA20–302A), 2020. *Ransomware Activity Targeting the Healthcare and Public Health Sector*. Cybersecurity & Infrastructure Security Agency. https://us-cert.cisa.gov/ncas/alerts/aa20-302a.

Chatelin, J., M.G. Stathopoulou, A.A. Arguinano, T. Xie, and S. Visvikis-Siest. 2017. Pharmacogenomic challenges in cardiovascular diseases: examples of drugs and considerations for future integration in clinical practice. *Current Pharmaceutical Biotechnology* 18 (3): 231–241.

Cohen, C. 1992. The case for presumed consent. *Transplantation Proceedings* 24 (5): 2168–2172.

Di Sanzo, M., L. Cipolloni, M. Borro, R. La Russa, A. Santurro, M. Scopetti, M. Simmaco, and P. Frati. 2017. Clinical applications of personalized medicine: a new paradigm and challenge. *Current Pharmaceutical Biotechnology* 18 (3): 194–203.

Emam, K.E., F.K. Dankar, R. Vaillancourt, T. Roffey, and M. Lysyk. 2009. Evaluating the Risk of Re-identification of Patients from Hospital Prescription Records. *The Canadian Journal of Hospital Pharmacy* 62 (4): 307–319.

Ghosh S. (2012). *Identity, Invention, and the Culture of Personalized Medicine Patenting*, 41 f. *et pass*. Cambridge University Press.

Gottlieb, S. 2003. Patients are at risk because of nurses' long hours, says report. *BMJ* 327 (7424): 1128.

Hardwig, J. 1990. What about the family? *Hastings Center Report* 20 (2): 5–10.

Hills, L. 2015. Managing the chronically overworked team: twenty strategies. *The Journal of Medical Practice Management* 30 (6): 388–392.

MacIntyre, A. 1981. *After Virtue: A Study in Moral Theory*. London, UK: Duckworth.

Miller, F.G., and H. Brody. 2003. A critique of clinical equipoise: therapeutic misconceptions in the ethics of clinical trials. *Hastings Center Report* 33 (3): 19–28.

Nyang, DaeHun, Aziz Mohaisen, and Jeonil Kang. 2014. Keylogging-resistant visual authentication protocols. *IEEE Transactions on Mobile Computing*. 13 (11): 2566–2579.

Porter, C.C. 2008. De-identified data and third party data mining: the risk of re-identification of personal information. *Washington Journal of Law, Technology & Arts* 5 (1): 1–8.

Ribaric, S., A. Ariyaeeinia, and N. Pavesic. 2016. De-identification for privacy protection in multimedia content: a survey. *Signal Processing: Image Communication*. 47: 131–151.

Richardson, V., S. Milam, and D. Chrysler. 2015. Is sharing de-identified data legal? the state of public health confidentiality laws and their interplay with statistical disclosure limitation techniques. *The Journal of Law, Medicine & Ethics* 43 (1_suppl): 83–86.

Rivers, C., B. Lewis, and S. Young. 2013. Detecting the determinants of health in social media. *Online Journal of Public Health Informatics* 5 (1): e161.

Sagiroglu, S., and G. Canbek. 2009. Keyloggers increasing threats to computer security and privacy. *IEEE Technology and Society Magazine* 28 (3): 10–17.

Sandel, M. 1982. *Liberalism and the Limits of Justice*. Cambridge University Press.

Sobhani, M., and L. Saxon. 2019. HIPAA Isn't Enough: All Our Data is Health Data. *USC Center for Body Computing*. https://medium.com/@usccbc/all-our-data-is-health-data-57d3cf0f336d. Accessed 07 Jan 2021

Wen, J., Y. Cheng, X. Hu, P. Yuan, T. Hao, and Y. Shi. 2016. Workload, burnout, and medical mistakes among physicians in China: A cross-sectional study. *Bioscience Trends* 10 (1): 27–33.

Woolhandler, S., T. Campbell, and D.U. Himmelstein. 2004. Health care administration in the United States and Canada: Micromanagement, macro costs. *International Journal of Health Services* 34 (1): 65–78.

Chapter 8
Members of Society

Abstract The present chapter examines the implications that the right to health care has for members of society. It outlines some of the limitations of this right, and identifies conditions under which the right turns into a duty. Matters such as informed consent are addressed from the perspective of patients, as are substitute decision making, privacy and access to research. The issue of addiction is also considered.

Keywords Addiction · Consent · Substitute decision making · Privacy · Research

8.1 Introduction

If the analysis in the preceding chapters is cogent then, ethically speaking, there is a right to health care, it is ultimately grounded in the Principle of Equality, and it has different albeit overlapping implications for the parties who are engaged in providing and delivering that care. More specifically, then, an ethical society has an obligation to institute an overall system that delivers health care to its members on an equitable basis; a corresponding set of obligations redounds to health care professionals and to health care institutions as purveyors of these health care services at the hands-on level, and these latter obligations can be seen as being, in a sense, derivatives of the overarching societal obligation. Finally, it was suggested that all of this should be consistent with the fundamental principles of ethics.

However, as was also suggested, the right to health care is not like a one-way street. More specifically, it was suggested that the parties that organize and deliver health care also have certain rights, and that neither these rights nor the corresponding obligations are absolute in the sense that they have no limits and do not come without conditions. Some of the implications of this were already spelled out in the preceding chapters. However, that was from the perspective of society, professionals and institutions. The purpose of the present chapter is to clarify some of these implications insofar as the right-holders—which is to say, insofar as the members of society—are concerned.

That, of course, is not all of what this chapter is all about. While the right to health care that members of an ethical society have as well as some of their duties were already discussed in the preceding chapters, they were essentially discussed from

E.-H. W. Kluge, *The Right to Health Care: Ethical Considerations*, The International Library of Bioethics 92, https://doi.org/10.1007/978-3-030-93838-3_8

the perspective of the service providers, and in many cases it was discussed only in an abbreviated fashion and deserves further analysis. Consent, is one of these issues, as is the matter of personalised medicine or the extent of the right to health care itself when the need for such care is brought about by addiction. The purpose of this chapter, then, is to fill in some of these gaps.

8.2 Limitations of the Right to Health Care

It may be appropriate to begin with a point that there are definite limits to the right to health care: limits that not only are grounded in the Principle of Impossibility but also in the right to health care itself.

The first factor needs little comment. As was pointed out before on several occasions, one cannot have an obligation to do what is impossible in principle or under the circumstances that obtain. Consequently the right to health care—and the corresponding the duty to provide such care—is limited by the resources that are available to society, where this includes not only resources in the sense of the material means that are necessary to provide such care but also the professionals who are needed to provide the relevant services, as well as the modalities that would have to be used when engaging in the relevant actions. When these are missing, the right to health care lapses.

8.2.1 Societal Considerations

One of the immediate consequences of this is that even though all ethical societies have an obligation to provide health care for their members, this does not mean that if a particular society does not provide all of the health care services that in principle would be implicated by the right itself or that are available in other societies, then that society fails in its duty and its member could legitimately claim that their right to health care was not being met. To argue thus would be to ignore the fact that providing health care is resource intensive. Therefore if a society's resource are insufficient to defray the costs of providing all possible health care services that are provided in other societies—or, indeed, if there is not a sufficient number of appropriately qualified professionals to deliver the services in question[1]—the Principle of Impossibility applies. This means that how the right of health care can legitimately be interpreted has to be adjusted in keeping with the facts of the situation.

Furthermore, as was also suggested, geographic conditions as well as demographic differences make distinct demands on the resources that are available to a society. For instance, a society that is geographically extended and covers diverse environmental and climatic conditions has to deal with different problems than a society whose

[1] Dussault and Franceschini (2006).

geographic footprint is small[2]; and a society with an extended age distribution faces different issues than a society whose population tends more to the younger end of the age spectrum.[3] Moreover, different societies face different challenges with respect to the other kinds of services that they are ethically obliged to provide because of the distinct genetic profiles of their populations.[4] This means that a given society cannot necessarily be ethically faulted when it does not apportion the same percentage of its overall resources to health care as some other societies. The percentile demand on its resources to satisfy the relevant obligations, as well as the gross national product that has to be divided among its various obligations, may simply be different.[5]

Picking up on the last comment, and to expand on this for a moment, an important fact that is often overlooked is that the resources that are available to a society cannot all be dedicated to providing health care. As was pointed out in Chap. 3, health care is only one of several obligations that belong to a society. Among other things, security, education and infrastructure etc. all have to be provided for, and all have legitimate claims on the resources that are available. These may differ simply because of differences in the factual embedding of the societies themselves, which means that different proportional allocations may be required to satisfy them. This, in turn, may limit the types and quantities of health services that the members of society can legitimately claim as being entailed by their right to health care, which is why the relevant claims may legitimately differ from society to society.

8.2.2 *Types of Care*

There is also the fact that what types of health are available even in principle is also determined by what may loosely be referred to as the state of the art. Not all health care modalities are at equal stages of development. Some are well established and effective and are appropriate for their specific range of application as this is recognised by duly constituted and recognized professional authorities, whereas other modalities lack that standing and are in the process of being researched or developed. As was pointed out in the preceding chapters, while societies have an obligation to engage in research in the case of health care modalities that fall into the latter rubric, this does not mean that these research modalities fall within the domain of health care services that the members of society can legitimately claim as their right.

This brings into focus the question whether access to research is a matter of right. To put it bluntly, while an ethical society has the obligation to engage in health care research, this does not mean that its individual members have the right to be included

[2] Sheiner (2014).

[3] Papanicolas et al. (2020).

[4] Di Rienzo (2006).

[5] WHO: *The Global Health |Observatory*, "Total expenditure on health as a percentage of gross domestic product", https://www.who.int/data/gho/data/indicators/indicator-details/GHO/total-expenditure-on-health-as-a-percentage-of-gross-domestic-product.

in such research. This is grounded in the nature of the right to health care itself as well as in the nature of scientifically valid research.

As to the first, since they are not established health care modalities, they cannot legitimately be considered as fulfilling the equity criterion that underpins the right to health care itself. At best, they hold out the possibility of meeting that criterion. However, the mere possibility of doing so does not entail that in fact they do—which in turn removes them from the range of modalities that members of a society can legitimately claim to be included in the array of health care services that the society is obligated to offer.

As to the second, a health care research project will be scientifically valid only if it meets proper statistical conditions, which in turn means that subject participation in such projects has to be strictly controlled lest the results do not reflect what actually is the case.[6] Therefore, again, members of society cannot legitimately claim that they have a right of access to research projects because their particular inclusion may invalidate the statistical profile of the relevant piece of research.

Furthermore, there are many different types of interventions that fall under the rubric of health care. They include such vastly different things as primary and secondary care, emergency care, mental health care, public health care, etc., but also interventions whose purpose is to satisfy a particular aesthetic perspective—what are usually called cosmetic interventions—and interventions that are intended to increase an individual's ability to compete at an increased level in professional sporting events. What includes these latter under the rubric of health care is the fact that in one way or another they affect the health status of their recipients. However—and this addresses the issue of limitation—members of society cannot legitimately claim such services as a matter of right. Ethically, members of society can only claim the kinds of health care services that satisfy the equity criterion that was discussed in Chap. 3.

The question that may arise at this point is whether such things as long-term care—care that is designed to meet someone's personal care needs so that they can live as independently and safely as possible when they can no longer perform activities-of daily living without assistance—also falls under the rubric of the kind of health care that members of society may legitimately claim. Moreover, and with due alteration of detail, the same question may be asked about such things as infant care, palliative care, etc.

While the discussion so far has proceeded as though these were included under the rubric of health care that the members of society may legitimately claim simply because in one way or another they involve health care modalities and interventions, the issue was never really addressed. Moreover, at first glance it seems unclear that they meet the equity criterion that ultimately determines what should and what should not be included under the rubric of right-to-health-care.

Closer consideration, however, shows that their inclusion may well satisfy that criterion. To see why, it is important to keep in mind that the criterion itself—which stipulates that the right to health care entails that society is ethically obligated

---

[6] World Health Organization (2020.

to provide health related services that would allow its members to take equitable advantage of the opportunities that are available in society—does not entail that all of the opportunities that are available in a society are of a competitive sort. Also included are such things as lifestyle, engaging in activities of daily living and the ability of self-expression without suffering from health-based impediments. Furthermore, these factors are not uniformly distributed on a longitudinal basis throughout the entire life-span of human beings but vary in accordance with such things as an individual's stage of development. Nevertheless, there is a fairly standard level for each of these that the members of society can reasonably expect to attain. It is this level that would appropriately be included among the opportunities that society offers its members, and it is with respect to these that the equity criterion may legitimately be applied, wherefore even in their case health care services would be implicated. Therefore they do fall under the rubric of services that members of society may legitimately claim as their right and that society is obligated to provide, while such things as cosmetic interventions that are based on individual aesthetic values and interventions that are designed to improve the ability to participate in professional sporting competitions and the like do not.

8.3 Person-Based Considerations

8.3.1 Responsibility

The last comments may serve as a transition from society-based to person-based factors that limit an individual's right to health care.

As was already pointed out in previous chapters, there are definite limitations and conditions that apply to the right to health care as it belongs to members of society when personal acts—or failures to act—on part of the members of society are involved. Thus, to pick up on a theme that was discussed under the rubric of allocation when this involves a competition between auto-induced as opposed to hetero-induced conditions,[7] when members of society voluntarily engage in activities that lower their health status and that lead to conditions that would otherwise satisfy the equity criterion and would in principle require health care interventions, this lowers their right to health care in certain respects. Even though the conditions may satisfy the equity criterion when considered on their own and in and by themselves, the fact that what led to these conditions lay under the individual's control puts them into the class of actions that violate the Principle of Nonmaleficence.

Differently, to voluntarily engage in an activity that results in a health condition that *prima facie* requires health care intervention would, if that intervention were provided, diminish the resources that would otherwise be available to provide health care for another person. However, the inability to provide the care that would

[7] Chapter 4, pp. 74 ff.

otherwise be available to that other person would result in avoidable harm for that other person. Consequently, by engaging in a type of activity that results in such an auto-induced condition, the individual would be the cause of harm to the other person and hence would be violating the Principle of Nonmaleficence. However, as was also pointed out in Chap. 3, the right to health care is a right that exists only within the framework of fundamental ethical principles, among which the Principle of Nonmaleficence is numbered. Therefore a violation of that Principle would restructure the strength of an individual's right to health care and may even entail that it be held in abeyance when selective choices have to be made, depending on the situation.

Perhaps it is worth noting explicitly that, with due alteration of detail, the same reasoning applies to failing to take appropriate steps to prevent the need for health care. The reasoning here centres in the nature of what it is to be the cause of a given state of affairs. Something may be the cause of a particular state of affairs either by actively intervening in the established course of events and thereby leading to a particular outcome, or by failing to intervene in the established course of events when it was possible to do so and thereby leading to another outcome. In other words, the notion of causality should not be understood simply in terms of exerting a particular effort but in being the determining factor for a given outcome.

The following example may illustrate the point at issue: If a patient in hospital experiences an acute cardiac event which, unless appropriately addressed, will lead to the death of the patient, and if that event is observed by an attending health care professional then, purely logically speaking, the professional has two options: either to intervene and engage in the appropriate intervention, or to stand by and merely observe the cardiac event unfolding. Each of these options leads to a certain outcome. On the first alternative the outcome, if successful, will lead to the patient not dying as a result of the cardiac event; on the second alternative the outcome will be the death of the patient because no appropriate action was taken. In both cases, however, the professional would have affected the unfolding of the course of events in a particular way. In other words, in each case the professional would have been the cause of the particular outcome: even though in the second case no effort or active intervention took place.

Returning, then, to the main reasoning, logically speaking the right to health care is embedded in a framework of ethical principles. These principles apply both to what is done and what is not done. Therefore if something is not done—in this case, failure to take appropriate steps to safeguard one's health and consequently making it necessary to access the limited health care resources that are available—then one would be causally responsible for requiring health care interventions and thereby diminishing the available pool of resources. This would be to cause harm to those who cannot have the interventions they require because the resources that are necessary for providing them have now become unavailable because they have already been spent. As was said, that would engage the Principle of Nonmaleficence. By that token, therefore, the strength of the right to health care of the individual who failed to take the appropriate steps to safeguard their health where such steps would be available to them would be reduced through their lack of appropriate action.

To be sure, it may be psychologically difficult for someone to engage in the appropriate actions that they know—or should now— would safeguard their health even though the means for doing so are available to them. However, psychological difficulty is not the measure of ethical legitimacy. If that were the case, then lying in situations where telling the truth would be psychologically difficult for the truth-teller would ethically legitimate their lying—which is not the case. Even the psychological difficulty that may be associated with not following what is a politically correct position is not an ethically legitimate excuse for not engaging in what otherwise would be ethically appropriate conduct. If that were the case, then not rejecting discrimination on the basis of gender, sex or ethnic extraction—which is politically correct in some countries in Asia[8] and in the Islamic world[9]—would also be ethically correct.

The overriding conditions for responsibility, therefore, are knowledge and ability: i.e., being aware of the options, and being in a position to do something about them. If the relevant information cannot reasonably be known by the individual—for instance, if an individual lives in an isolated community and could not reasonably be expected to know that not engaging in some form of exercise will lead to health problems—then that would be an ethically acceptable excuse for not doing so; or if an individual was unable to perform the otherwise appropriate action—for instance, if the individual was a quadriplegic and therefore could not engage in cardio-vascular exercise—then this would also be an ethically legitimate excuse for not acting. However, when exculpating conditions such as these are not met, the individual's obligation to engage in the relevant action remains simply because they are persons among other persons.

8.3.2 *The Issue of Personhood—Again*

The last comment brings into focus an issue that was already touched on in the preceding chapters: namely, the issue of personhood. The focus in those chapters, however, was that there are culturally based differences in how the notion of personhood may be understood—the atomic as opposed to the embedded conceptions were used as illustrations—and that this had distinct implications with respect to the issue of informed consent. By contrast, in the present context the issue is whether distinct construals of the notion of personhood irrespective of the atomic versus embedded conception have different implications for to the right to health care itself.

More specifically, and as was said before, one fairly common historical construal of the notion of personhood is that being a living member of the species *homo sapiens* is a sufficient condition for being considered a person. On this conception of personhood it follows that an individual who satisfies this criterion has a right to health care irrespective of whether they exhibit other characteristics. In other words, even though the individual may in fact be whole-brain dead—i.e., may have an

[8] Mehrotra and Kapoor (2009).

[9] Al-Ahsan (2009).

irreversibly non-functioning brain inclusive of the brainstem[10]—being biologically alive and meeting the biological criterion would be sufficient to justify counting the individual as a person and claiming that they have a right to health care.

By contrast, the currently more generally accepted notion of personhood—certainly in in health care circles—is that of a living member of the species *home sapiens* who has a functioning brain inclusive of the brainstem. The individual may not be conscious, may not be sophisticated or capable of making decisions in any rational manner—newborn infants, unconscious individuals as well as those who suffer from serious psychiatric issues, etc. would here be included—as long as they are not whole-brain dead they would still count as persons. This would be the case even if they did not meet the cardio-pulmonary criterion, as would be the case with someone who was artificially ventilated and whose cardiac function was taken over by a machine. Therefore on this conception of personhood only individuals who are not brain-dead in the above sense of the term would count as persons and would have a right to health care

Clearly, then, how the notion of personhood is interpreted has important implications for how the limits of the right to health care are drawn from the perspective of patients. Historically, the issue was brought to the fore in 1968 with the publication of "A Definition of Irreversible Coma" by the Ad Hoc Committee of the Hard Medical School to Examine the Definition of Death.[11] Essentially, the Committee advocated a whole-brain criterion. A basic premise that underlay the Committee's reasoning was that an organ that no longer functions and has no possibility of functioning again is for all practical purposes dead. Another basic premise to which the Committee subscribed was that the defining characteristic of human beings as persons was not simply that of being a functioning biological entity but of being a functioning biological entity that has the capacity for reasoning, self-awareness and decision-making. On this basis it concluded that an individual whose heart continues to beat but whose brain inclusive of the brainstem was irreversibly non-functioning would still be a human being in the biological sense of that term but would no longer be a human person in an ethically relevant sense. In presenting its reasoning, the Committee spelled out several physiological criteria that would have to be met by an individual in order to cease being considered a person—criteria that have since been refined in several ways[12]—and these were compendiously referred to as criteria for brain death. It is this notion of brain death that has variously been recognized as definitive of personhood in many jurisdictions, although some of them have made it mandatory that rather than the brain death criterion being automatically applied, the patient's antecedent wishes regarding the use of the brain death criterion be considered decisive.[13]

Plainly, then, the issue of which conception of death—and, correspondingly, which conception of personhood—is ethically defensible has important implications

---

[10] Austriaco (2016).

[11] Ad Hoc Committee of the Harvard Medical School to Examine the Definition of Death (1968).

[12] Drake et al. (2017).

[13] Cohena et al. (2012).

for how the limits of the right to health care should be drawn. On the one hand, if the traditional cardio-pulmonary criterion is considered to be determinative, then an individual whose heart is beating and who is breathing (even though with assistance) would still be a person and therefore have a *prima facie* right to health care. On the other hand, if the individual met the modified Harvard Criteria and therefore was declared brain-dead, then the individual would no longer count as a person, which in turn would mean that the individual's right to health care would be extinguished.

8.3.2.1 Ethics Versus Ethos

It is at this juncture that the distinction between ethics and ethos—i.e., between ethics and a system of traditionally accepted values and beliefs that is espoused by a person, a group of persons, or by an institution—becomes relevant once again.

As was suggested in Chaps. 1 and 2, ethics is grounded in fundamental principles that determine whether a given action or state of affairs is morally acceptable. As such, it is independent of a particular individual's or group of individuals' beliefs, and it is independent of cultural or institutional values. Further, it construes the notion of a right as a justified claim that someone is entitled to raise if they wish to do so, and it interprets the notion of an obligation as its correlative.

Applying this to the question of how the difference between ethics and ethos affects the right to health care, it follows that if ethics is the determinant, then only entities that are in principle capable of making a claim can have ethical rights and that therefore only they have a right to health care. While they may not be able to exercise their rights on their own behalf, either in general or on a given occasion—they may be unconscious, lack sufficient sophistication, be too young to engage in decision-making, etc.—if they meet this criterion they will still count as persons and therefore will still have rights. Consequently—to apply this to the issue of brain-death and of being purely biologically alive—if they are brain-dead and lack a functioning brain inclusive of brainstem[14] and hence cannot even in principle make claims, then they cannot be considered to be persons, cannot be the subjects of ethical consideration and therefore cannot have a right to health care. They may be legally entitled to such care, but that is another issue.

That conclusion, however, follows only on the basis of ethics. By contrast, on the basis of ethos—i.e., on the basis of a system of traditionally accepted values and beliefs that are espoused by a person, a group of persons or by an institution—this would not follow. An ethos may construe the notion of personhood in any way it wishes. Therefore a particular ethos may very well reject the brain-death criterion and maintain that being biologically alive and a member of the species *homo sapiens* are themselves sufficient to ground an ascription of personhood. Consequently, on such an understanding, brain-dead individuals who still have cardiac and pulmonary function would still be persons and therefore have a right to health care.

---

[14] Or its equivalent. It is currently unclear whether the neurological structures of organisms like octopuses qualify as functional equivalents of brains in the sense under discussion.

However, if the reasoning above as well as in Chaps. 1, 2, and 3 is cogent, then ethically speaking the right to health care belongs only to persons in the sense of individuals who are not brain-dead as this was outlined by the modified Harvard Criteria. This, then, means that even though the next-of-kin or significant others of a given individual may subscribe to an ethos that defines personhood in terms of cardio-pulmonary criteria and therefore maintain that although the individual is brain-dead they are still a person and hence should still receive health care, that would be an expression of their personal perspective and would not be ethically binding.

The question, of course, is whether an ethics-based or an ethos-based reasoning should define the right to health care and its limits. The answer here depends on how the term "should" is understood. If it is understood in an ethical sense—i.e., in the sense of something being obligatory in light of the fundamental principle of ethics as these were outlined in Chap. 2—then the answer is that ethics-based reasoning should be decisive, which in turn means that the notion of brain-death should define the limits of an individual's right to health care. On the other hand, if "should" is understood in a sense that is grounded in ethos, then an individual's right to health care would not be drawn by ethical considerations but on the basis of tradition, belief and the like—and perhaps even on the basis of established religious, legal or political considerations.

8.4 Health Care Itself: Perspectival Considerations

A further question that arises relative to how the limits of the right to health care should ethically be drawn with respect to patients—and one that also has important implications for where the limits of the right to health care should ethically be understood at the policy level—centres in how the notion of health care itself should be construed.

The standard view of health care interventions is that this includes all and only those interventions where a causal relationship can be established between the intervention in question and its health related outcome; further, that the intervention must either improve the health status of the individual, maintain it or retard its negative progression in a causally demonstrable way. This, then, excludes interventions that rely on the placebo effect: which is to say that it excludes interventions whose outcome relies on the perception of those who evaluate it but where no demonstrable causal relationship exists between the intervention and the outcome itself. The placebo itself need not be chemically or biologically inert, but in order to be considered a placebo it would be sufficient that it not be causally efficacious for the condition in question.[15]

[15] This understanding of the term "placebo" is distinct from "active placebo" in research. The addition of the term "active" indicates that the intervention itself has causal efficacy, while the rem "placebo" indicates that it is not its causal efficacy that is being investigated. See Vase and Wartolowska (2019).

On the other hand, a non-standard view of how health interventions function may differ in various ways from what may thus be called the scientifically grounded conception. One example here would be homeopathy. Homeopathy contends that the human body can cure itself of illnesses when it is properly stimulated, where such stimulation involves the use tiny amounts of natural substances like plants or minerals and even of a tiny amount of the substance that causes the symptoms of a disease itself, and that applying these to the affected person will restore their health. It is based on the view that like causes like—hence the name of the approach itself—and the claim that the lower the dose of a given medication, the greater its effect: something that is known as the "law of minimum dose."[16]

Another example would be the views that compendiously may be referred to as alternative medicine: views that are culturally or valuationally grounded and that include traditional ethnic medicines according to which appropriate health practices include things that incorporate plant, animal and mineral substances as well as spiritual therapies, manual techniques and exercises, and that are used either on their own or in combination to treat, diagnose and prevent illnesses or to maintain well-being.[17]

In each case, of course, the question is what role such variations in the understanding of the nature of health care itself may legitimately play relative to an individual's claim to have the respective interventions included in the domain of interventions that are implicated by the right to health care.

If the analysis in the preceding chapters is cogent, then an individual's right to health care is not a right to what the individual considers effective but a right to what in fact is effective according to the best data that are available at that point in time. The reason for this lies in how, ethically, rights and obligations are established. They are not established on the basis of what individuals wish or want, nor even on the basis of what they believe, but on the basis of ethical principles. In this regard, once more, one of the crucial facts is that society's resources are limited. Another fact is that society has an obligation to provide health care to its members. A third fact is that if a society were to spend its resources on the basis of values or beliefs that did not have demonstrable causal validity, then that society would not be acting ethically even though their members were to feel that their beliefs and rights were being respected, because it would be wasting its limited resources on intervention that have no causally demonstrative effect. In doing so, it would violate the Principle of Nonmaleficence since it would be causing preventable harm by not having the resources to prevent health conditions because the necessary resources that would be necessary to do so had already been spent on inefficacious interventions.

From this it follows that even when the members of a society share a particular belief about what health care interventions are effective, and even when they construe the notion of effectiveness itself on the basis of value-based considerations or on the basis of tradition, this does not establish an ethical obligation on part of society, and does not mean that the members of society therefore have a right to the interventions

[16] Royal Pharmaceutical Society, "Homeopathy"; https://www.rpharms.com/resources/quick-reference-guides/homeopathy.

[17] White House Commission on Complementary and Alternative Medicine Policy (2002).

that they claim. It may be politically or even legally appropriate to do so. Ethically, however, that is another matter, for ethical rights and obligations arise only on the basis of facts as combined with ethical principles. As was suggested above, this entails that effectiveness in the scientifically grounded causal sense of that term is the determinant of which interventions may legitimately be included in the domain of obligatory health care.

8.5 Consent[18]

This brings into play two further ethical considerations: one that centres in the issue of informed consent, and another that centres in the matter of privacy.

The issue of consent is implicated because even though members of society may have a right to health care, all other things being equal—exceptions will be discussed a little later—whether they request that care or accept it when it is offered is a matter of informed consent on their part. To restate what was argued in greater detail in Chaps. 6 and 7, a right is not a duty, and whether a right is exercised is a matter of personal decision-making. As was then also pointed out, consent is ethically valid only if the consent-giver is properly informed about the nature of the intervention that is being offered, about its likely effects both positive and negative, and about how a refusal is likely to affect the health status of the individual in question. Moreover, the relevant information should be meaningful for the patient and therefore would function as the basis of a valid consent only if the individual who gives that consent has been offered the relevant information in a language and at the level that a reasonable person in their particular position would understand— and the individual does in fact understand it.

Clearly—and aside from the matter of what actually goes on in the process of consent itself—this establishes a connection between the amount and type of information that a patient has when making a decision and the validity of the decision that the patient ultimately makes. Simply put, the more accurate and understandable the information that the patient has and that is relevant to the decision that they are to make, the more informed their decision is likely to be. And here it is important to note that, purely logically speaking, it is irrelevant whether that information is provided by the health care professional who is seeking the patient's consent for a particular intervention or it stems from some other source. What is important is that the information be relevant and above all meaningful and accurate.[19] The role of the health care professional is to ensure that these requirements are in fact met by the consent-giver.

It is at this juncture, then, that the ethical obligations of society and of health care professionals relative to patient consent overlap with the rights of patients. To put

---

[18] For a discussion of informed consent from the perspective of health care professionals, see Chap. 6, pp. 120 ff.

[19] Delbaere and Smith (2006).

it bluntly, patients have the right to seek the relevant information from any source they wish, society has an obligation to take appropriate steps to ensure that the health care information that is provided to patients through such things as the public media is accurate and not misleading; and health care professionals have an obligation to ensure not only that the patients understand the information with which they are provided but also that the information is relevant, complete, accurate and applicable to their situation.

If that is correct, then in principle there is nothing ethically untoward about the information being accessed by patients through public media and advertising. While the reason why those who provide such information—pharmaceutical corporations who advertise their products would here be implicated, as would be private hospitals or health care facilities that advertise the surgical and other services that they offer for profit—may be different from the therapeutic intent of the health care professionals who seek their patients' consent for a given intervention, logically speaking that is irrelevant for the simple reason that it is irrelevant where the information comes from. As was said, what is relevant—indeed, what is crucial—is that the information be relevant, complete, accurate and applicable to the situation; and what is important about the process of giving consent itself is that the health care profession who seeks the patient's consent provide the relevant information and review the accuracy and applicability of the information that otherwise may be in the patient's possession.

Moreover, another important fact about consent is that it is, so to speak, a two-way affair. It involves both the health care professional and the patient. While the Principle of Autonomy entails that the right of informed consent belongs to the patient, and while it is the patient's decision whether to accept a particular prescription or agree to a particular intervention, the patient's consent is, so to speak, only empowering. It is the health care professional who ultimately decides whether to act on that consent. In other words, the health care professional functions as gate-keeper to health care; and if the professional is not convinced that the information on which the patient bases their decision is correct or relevant, or that the patient really understands the significance of that information then—again, emergency contexts possibly excepted—the professional is not obliged to act on that consent. That would follow only if the intervention fell under the professional's obligation to act in the best interest of the patient.

In this connection, it may be worth emphasizing that a patient's right to consent to health care does not entail that a patient has the right to insist that a particular intervention should be offered. To argue thus would be to confuse the right of consent—which is the right to give permission to do something that is being offered—with the right to demand, which is the right to insist that something be offered.

In this connection, there is a further point about consent that is of fundamental importance for appreciating what a patient's right to health care amounts to. To put it first in purely logical terms, there is a fundamental logical difference between the right to health care, the right to a type of health care and the right to a specific health care invention. The right to health care is the right to an array of modalities of health care interventions; the right to a type of health care is the right to a particular modality of health care intervention, and the right to a specific health care intervention is the right to an instance of such a modality.

Applying this distinction to the issue at stake, it follows that the right to health care itself is engaged when someone suffers from a health condition that prevents the individual from taking equitable advantage of the opportunities that society opens up for its members. Logically speaking, therefore, this means that if someone suffers from something like breast cancer or prostate cancer, then that individual has the right to types of intervention whose purpose is to address that cancer in a way that satisfies the equity criterion. However, this is a right that spans a series of relevant health care modalities because there are various types of interventions that deal with these conditions. Therefore having the right to health care does not mean that the individual has the right to a particular kind of cancer treatment. All it means is that the individual has the right to the kind of treatment that is most likely to deal with the cancer in an effective manner. Nor does it mean that the patient has the right to a specific way of engaging in the particular kind of treatment. While a health care professional may offer the choice between such options to the patient, the specific way of engaging in the options that are offered to the patient is a matter of professional decision.

The last comment may also benefit from more detailed discussion, because it also addresses the issue of limitation relative to patients' right to health care. As was said above, what ultimately validates the comment itself is the fact that the resources that society has its disposal to satisfy its obligation to provide health care are limited. In accordance with the Principle of Nonmaleficence, this means that society must use its health care resources in a way that does not either produce harm or increase the likelihood of harm resulting from its actions; and this, in turn, means that it must use the resources it has in the most effective way possible. It therefore follows that if, for a given health condition, there are various possible treatment modalities that would deal with that condition, then the Principle of Nonmaleficence entails that society is only ethically obligated to provide the kind of intervention that is most likely to effectively deal with the condition. Whether it decides to offer more than that is a matter of politics or tradition.

Of course, whether a particular type of intervention will be effective in a given case is never certain. Whether something will be effective depends on the precise nature of the condition as well as on the overall health status of the individual in question. That is why the options that are outlined to a patient for their consent, when they are presented properly, are not presented in definitive terms but in terms of probabilities. Therefore part of the role of health care professionals in each case is to estimate which type of intervention is most likely to be the most effective for the particular condition of the particular patient. The preferences of the patient here play no real role in this—except, perhaps in the sense that if the intervention that is proposed by the professional is the sort of intervention that agrees with the patient's preferences, the patient is more likely to follow it. That, however, does not establish a duty on part of the health care professional do accede to the preference-based wishes of the patient. The professional's obligation to provide health care is satisfied when the most effective kind of intervention is offered for the patient's consent.

8.6 Substitute Decision-Making

Before turning to the issue of privacy, it may be appropriate to consider briefly the matter of consent in the context of individuals who, for various reasons, cannot give consent on their own behalf. While the matter was discussed in some detail in Chap. 6,[20] that was from the perspective of health care professionals, whereas the context here is that of the patients themselves.

It is generally agreed that the right to health care does not lapse because of an inability to consent to a health care intervention nor, by the same token, does the right to refuse consent lapse because of an inability to indicate the refusal in an appropriate manner. Rather, as was pointed out before and as based on the Principle of Equality, society has the obligation to put in place an effective mechanism to identify someone who will make the decision on behalf of the affected person lest the right to health care lapses. This means that patients have the right to expect that society will enact the appropriate provisions.

However, both who counts as someone who requires a substitute decision-maker and who that decision-maker should be is not universally agreed, and is a matter of some controversy. For instance, in some jurisdictions individuals below a certain age are deemed to be incompetent to exercise their right to give consent,[21] whereas in other jurisdictions that is a rebuttable presumption and what is considered to be telling is the individual's cognitive and volitional capacity.[22] Likewise, who should act in the role of substitute decision-maker is legally determined in some jurisdictions on the basis of the consanguineal or analogous relationship in which the individual stands to the person in question, whereas in others it is determined on the basis of the social relationship in which the two stand to each other, where this may include both of the preceding as well as partnership relationships and the like. Finally, there is the matter of what values should guide the substitute decision-makers: Should it be their own values, the values of the social context in which the incompetent person is embedded, or some other considerations like that of the likely effectiveness of the proposed intervention?

Beginning with the question of who should assume the role of substitute decision-maker, ethically speaking the answer does not depend on the relationship between the incompetent person and the proposed decision-maker, however that relationship may be construed. That would be the case if what was at stake was a matter of relationships. In that case it could be settled on the basis of the nature of the relationship itself and of the propinquity that obtains between the two parties.

However, substitute decision-making is not a matter of relationships—although, as will be clarified in a moment, that may have a bearing on the issue—but a matter of exercising the decision-making right of the incompetent person in an ethically appropriate manner. Therefore the fact that the incompetent individual and the proposed

---

[20] Chapter 6, pp. 124 f.

[21] Cf. Japan, Article 4 (1) of the *Child Welfare Act*. See also McNary (2014) for US variations.

[22] Cf. [1985] UKHL 7, [1986] AC 112, [1986] 1 FLR 229, [1985] 3 WLR 830; *A.C. v. Manitoba (Director of Child and Family Services*, 2009 SCC 30.

decision-maker stand in some sort to relationship to each other is essentially irrelevant. What is ethically central is the substitute decision-maker's ability to employ the appropriate values when making the decision. Otherwise, while the decision would be made in the name of the incompetent person, it would not be the decision that the incompetent person would have made if they had been able to make the decision on their own behalf. And while relational propinquity may be an indicator of that ability, the test is whether that in fact is the case.

In this connection, it may be relevant to point out that most jurisdictions acknowledge this point, because they allow a proposed substitute decision-maker to be legally challenged if there are reasonable grounds to suppose that they will not do so or are not be able to do so: for instance, if they have been out of touch with the incompetent person for a certain period of time, or if there are reasonable grounds to suppose that the values that they would use would not be the appropriate values. To reiterate, then, irrespective of the relationship in which the incompetent person and the substitute decision-maker stand to each other, the central question is whether the substitute decision-maker would use the appropriate values when making the decision on behalf of the incompetent person.

This, of course, entrains the question of what these values should be. Three possibilities were hinted at above: the substitute's own values, those of the social context in which the incompetent person is embedded, and the value of effectiveness.

For obvious reasons, the suggestion that it should be the substitute decision-maker's own values is most likely to be ethically unacceptable: Unless there are reasonable grounds to suppose that the incompetent person would have the same values as those of the substitute decision-maker, the substitute would not be exercising the incompetent person's decision-making right but would be arrogating that right to themselves. Of course there may be grounds to suppose that the two share—or would share—the same values. For instance, this may be the case with young children and their primary caretakers. However, even in this context there is the possibility that the primary caretakers' values may not be in keeping with the standard values of society in general: for instance, if the substitute decision-maker subscribed to Jehovah's Witness values or the values that shape the decision-making of members of the Church of Christ, Scientist. In these sorts of cases, to allow the primary caretakers to function as substitute decision-makers and use their own values may well be to deprive the incompetent persons of the right to health care that belongs to all members of society, and would do so simply because they stand in a particular relationship to their substitute decision-makers.

As to using the values of the specific social or ethnic sub-grouping in which the individual is embedded, that is also not necessarily without ethical difficulties relative to the right of the incompetent person. The particular social or ethnic sub-grouping in which the incompetent person is embedded may subscribe to values that are distinct from those of society in general. For instance, this would be the case if the incompetent person was socially embedded in a particular sub-cultural framework that, in distinction from the rest of society, construed the adherence to ethnic values as integral to decision-making and that rejected a scientifically grounded understanding of health and health care and instead subscribed to a valuational perspective that considered adherence to traditional or ethnic beliefs as determining. This would be the case if the incompetent individual was socially embedded in a First Nation that did not accept scientifically or "Western" medical practice and instead relied on traditional approaches to health care and insisted that these should be followed in all health-care decisions. If the substitute decision-maker followed these values when making a health care decision for the incompetent individual that would be to deprive the incompetent person of what in fact was their right as member of the overall society in which the First Nation itself was embedded. This would therefore be unethical and would in fact be discriminatory. To paraphrase an expression from a different context, it would be to assume that inheritance and embeddedness are destiny.

The last comment may benefit from further discussion. Ultimately, it is grounded in the logic that underlies the existence of a right itself as well as in the Principles of Autonomy and Equality. That is to say a right—and that includes the right to health care—does not exist unless there is someone whose right it is. Therefore unless the incompetent person for whom substitute decision-making is an issue is alive, the matter of substitute decision-making itself as well as the questions of who should be the substitute decision-maker and of what values they should use when acting in that capacity would not arise.

At this point the analysis splits into two streams: one that deals with persons who are incompetent and will never be competent, and another that deals with persons who are incompetent but will likely become competent.

In the first sort of case, if the substitute decision-maker were to ignore the value of effectiveness and instead use the traditional values of the social context in which the incompetent individual was embedded and if these values would lead to their not choosing an effective intervention then—to use the expression that was used before—that would be for the substitute decision-maker to decide that inheritance and embeddedness were destiny.[23] By proceeding in this fashion, however, they would treat the incompetent person as not having the same right to health care as other members of society who are not born into that sub-cultural grouping or embedding but are members of the overall society of which the sub-group was part. In other words, if following these traditional values would mean not deciding on the most effective and otherwise standard option that would underwrite the continued existence of the incompetent person, then the right to exist would take second place to adhering to the values and standards of the embedding sub-culture. However, while that may be a decision that a competent individual may make on their own behalf, it is not a decision that a substitute decision maker is entitled to make for someone else: in particular if the individual in question has never been and never will be competent. It would be to violate the Principle of Equality—after all, the Principle applies to individuals as persons, not to persons-as-embedded. It would also, and thereby, be to violate the Principle of Nonmaleficence.

[23] Cf. *Hamilton Health Sciences Corporation v. D.H.*, *McGill Journal of Law and Health* (2016), https://mjlh.mcgill.ca/2016/03/21/revisiting-hamilton-health-sciences-corporation-v-d-h/.

The second stream focuses on the fact that people are not born with ethically defensible and autonomously and competently chosen values. In order to have such values, they must not only exist but also develop and mature. Therefore when faced with a situation where health care is an issue for an incompetent individual who will become competent if they survive, then their substitute decision-makers are obligated to choose the option that is most likely to ensure the continued existence of the now-incompetent person so that they can develop and choose the values that they will use in their own decision-making when they are competent.

Applying these considerations, it means that the Principles of Equality and of Nonmaleficence ultimately entail that even a society in the sub-cultural sense of that term has an obligation to provide all those health services for its incompetent members that are the most effective and that are available in the overall society in which the sub-culture itself is embedded. In other words, ethically speaking this holds however the nature of society itself is construed—i.e., whether it is construed on a cultural, sub-cultural or legal basis—simply because it is the overall society that contains its various sub-groups and the individuals in question are also members of that overall society and are persons, their sub-cultural embedding notwithstanding. This, therefore, means that since ethical substitute decision-making should occur within the overall ethical framework that governs a society as a whole and not simply that of the particular sub-group of that society, a substitute decision-maker should use the value of effectiveness if that was the value that was standardly used in the case of individuals who are not embedded in the particular sub-group.

8.7 Privacy

Of course the question of what type of intervention will be effective is not obvious in itself, but is a matter of data analysis. That, however, is not merely a matter of analysing in abstract and purely scientific terms the data that are available relative to the nature of the intervention in the laboratory setting but of what happens in actual life. This, in turn, involves taking into account how the particular conditions and health status of individual patients are affected by a given intervention. Therefore the matter of the right to health care for individual patients branches out into the issue of research, because it is only in this way that such data can be established. And that, in turn, triggers the issue of privacy.

8.8 Research

To put this into context, it may be appropriate to begin once more with what happens when a health care professional offers an intervention to a patient. That offer is not a guarantee that if the patient receives that intervention, the intervention will be effective. Instead, it is an offer that is based on what may be called the probability profile of the intervention. That probability profile, as was discussed in the previous chapter, is established on the basis of data that are derived initially through Phase I-III research[24] and then, if appropriate, are modified on the basis of Phase IV studies that look at how the intervention performs in actual practice when the controls that are involved in Phase I-III trials are not in place.[25]

However, as was pointed out in Chap. 5, the actual scientific analysis of the effectiveness profile of a given intervention in real life involves engaging in research protocols that analyse the data that are contained in the records of patients who are not necessarily involved in Phase I-III studies. Therefore absent patient consent to make the relevant parts of their records available for such studies, the actual effectiveness profile of an intervention cannot be calculated. This means that absent such data-based studies, it would be impossible for a health care professional to suggest that a particular type of intervention would be appropriate for the particular condition of a given patient, and it would also be impossible for society to decide what kinds of interventions to fund as a matter of policy in order to fulfil its obligation to provide appropriate health care for its members.

It therefore follows by the Principle of Impossibility that if patients wish to exercise their right to health care, then they cannot insist on their right to privacy being strictly adhered to outside of the physician–patient relationship. More specifically, they must be willing to make the relevant aspects of their record available to duly qualified researchers who are engaged in scientifically valid and ethically sound statistical research that deals with the types of interventions when these are no longer delivered during strictly controlled Phase I-III studies. Further—as is legally recognised in some jurisdictions—it follows that they must permit such access without explicit consent on the condition that otherwise applicable privacy conditions not be breached and the researchers be bound by the condition of confidentiality.

8.9 Personalised Medicine

All of this seems to be essentially unproblematic in the context of health care research as it is traditionally understood. Traditional research patterns essentially deal with the health conditions as these are presented in particular cases and in terms of the outcomes of interventions in relation to the health status and conditions of the relevant patients. In these sorts of cases, permission to access their record, whether explicitly given or tacitly entailed, satisfies the privacy rights of the individual patient because it essentially involves only the privacy of the individuals whose records are being accessed and has no implications for the privacy rights of other parties.

---

[24] In some jurisdictions, as for instance the United States of America, Phase 0 trials are suggested in order to demine whether the pharmacokinetics and pharmacodynamics of a given pharmaceutical warrant moving ahead to Phase I-III trials.

[25] Suvarna (2010).

In the case of research into personalised medicine, however, the situation appears to be different because of the genetic component of the research. More specifically, it seems that the use of the genetic data, which is a central feature of personalised medicine, may allow the identification of other individuals who share the same genetic features and may therefore constitute what in effect would be an unconsented-to breach of privacy.

It may be useful to illustrate what is here at issue by considering an example. The incidence of Tay-Sachs disease—which is an autosomal recessive condition that is caused by mutations in the HEXA or beta-hexosaminidase A gene—is higher among Jews of Ashkenazi descent than among other populations.[26] Therefore, for an individual who is a carrier of Tay-Sachs disease to allow the genetic data in their record to be accessed in order to develop a treatment for that condition would thereby appear to identify without their consent their consanguineal relatives, and indirectly the whole group of persons of Ashkenazi descent. Hence it would appear to violate their right to privacy. The question, therefore, is whether this is ethically defensible, and therefore whether the development of personalised medicine which makes use of these and other genetic data that directly or indirectly identify other individuals runs into a problem of privacy.

At first glance, it would seem that this is indeed the case: i.e., that for an individual patient to allow such access to their record inclusive of their genetic profile amounts to a violation of the right of privacy of these other persons because the inclusion of the genetic data seems to allow the identification of these other persons. To allow that to happen, however, would be to hand over to the individual patient the right to control the privacy rights of the other persons.

However, the problem may be more apparent than real. In the first instance, as to the genetic data, while they are shared among consanguineal relatives and even among groups of ethnically related persons, this does not mean that the data themselves carry identifiers with them, which would be required if there was a threat of an unconsented-to privacy invasion. As forensic protocols have shown, genetic data by themselves cannot be used to identify anyone unless they are linked to a larger set of personal data in which they are included, and it is the latter that provide the identification.[27] Logically speaking, this is no different from the sort of situation where a correlation is established between a hydroxyl radical and a reaction to a particular chemical. Establishing that relationship does not in itself identify all substances that share the hydroxyl radical. It merely establishes the type and nature of the reaction itself. Therefore while access to the genetic data that are involved in developing personalised medicine may potentially serve to identify individuals or groups of individuals, it does not in itself do so. As was said, that would be the case only if the genetic data carried personal identifiers with them, and that is not the case.

Furthermore, as was pointed out before, if members of society wish to exercise their right to health care, then they must make it possible for society to provide

[26] "Tay Sachs Disease". NORD (National Organization for Rare Disorders). 2017, accessed 08/02/2021 at https://rarediseases.org/rare-diseases/tay-sachs-disease/.

[27] Panneerchelvam and Norazmi (2003).

such care. Personalised medicine allows—or at least promises to allow—society to do so in a most effective and efficient manner. Not only that: It also maximizes—or promises to maximize—the ability of health care professionals to at in the best interests of their patients.

More particularly still, personalised medicine is predicated on the thesis that how—and, indeed, whether—a given patient reacts to a specific intervention (and, in particular, to a pharmaceutical medication) is functionally related to the patient's genetic profile because it is the patient's genetic makeup that ultimately steers the patient's biological processes. Therefore by correlating the outcomes of particular types of interventions with the genetic profiles of the individuals who have received them, it is possible to determine the effectiveness profiles of these interventions more precisely than simply by proceeding on the basis of statistical analyses of the data that are assembled when engaging in more or less traditional research and practice patterns. Consequently, by researching into and developing methods of personalised medicine the patients' right to health care would be satisfied more effectively, and society's obligation to provide health care could be met more efficiently, and would extend the health care professionals' ability to act in the best interest of their patients.

However, research into personalised medicine is not generally conducted in Phase II-III trials, which involve hands-on interactions between researchers and patients. It often—in fact, it generally—involves the analysis of data that are contained in patient records that are stored in data banks, and this is where some of the considerations that were raised about so-called second-hand research in Chap. 5 become relevant. Such research involves what is sometimes called data mining: i.e., the use of informatic techniques to scan for relevant factors that are contained in the records that are stored in these data banks.[28]

The privacy issues that arise with respect to data banks were already discussed to some degree in Chaps. 5 and 7. However, it may be worth restating first what these amount to in general terms, because health records are not always maintained by a health care institution, and then focusing on the alleged privacy issue that would be involved in data mining in the context of research in personalised medicine.

When patient records are stored in larger data banks, it is generally required that access to them be restricted to properly placed and qualified individuals who are engaged in ethically and scientifically valid and certified projects, and who are bound to abide by valid rules of confidentiality. A good example of this is the provisions that were put in place by the European Union's International Consortium for Personalised Medicine. In its General Data Protection Regulations (GDPR) the Consortium stated that access to clinical and population-based biobanks was necessary in order to engage in effective research and for the development and application of personalised medicine, but it stipulated that such access should only be permitted in compliance with relevant data protection legislation, and that this should be the case even if such research was conducted on an interjurisdictional basis.[29]

[28] Chen and Fawcett (2016), Ngiam and Khor (2019).

[29] European Council Conclusion on Personalised Medicine for Patients (2015/C 421/03), accessed 04/02/2021 at https://www.icpermed.eu/en/icpermed-medicine.php.

Arguably, of course, even though such provisions as the GDPR have an ethical basis,[30] they do not directly address the ethical issue of privacy that appears to be involved in research that is specific to personalised medicine. However—and this takes the discussion back to the issue that was identified a few paragraphs ago—the assumption that specific ethical issues arise in this regard is itself based on the assumption that the genetic data that are used in developing personalised medicine are sufficient to identify particular groups in terms of ethnicity, and that they allow the identification of specific individuals as belonging to those particular groups.

Following up on what was said above, however, that assumption is highly questionable. Thus—and this takes the discussion back to what was said a moment ago about genetic data and identification—the mere fact that a particular genetic factor is correlated with a particular condition does not in itself identify a group or an individual. That would be the case if and only if the genetic factor was unique to that group or that particular individual; and even then it would require independent identification of the group itself or of the individual by some other means. Therefore the right to privacy could not be breached on that basis.

As to data mining, the ethically important fact is that it involves informatic tools, and that it does not involve interaction between the researchers who use the tools and the patients as persons. Consequently, if access to the patient records is limited to protocols whose purpose is to develop or deepen an understanding of the correlation between particular genetic conditions and particular health care interventions, and if the protocols themselves have been validated by a duly constituted ethical authority; and if, furthermore, the users of these protocols have themselves been approved by such a authority, then data mining in personalised medicine research does not present any special ethical issues.

To sum, up, then: By the Principle of Impossibility, the right that members of society have to health care entails that society has a corresponding obligation to provide such care to the best of its ability. This means that society has an obligation to develop personalised medicine techniques in order to meet its obligation and so that health care professionals will be able to fulfil their obligation to act in the best interest of their patients. All this, however, is possible only if patients agree to have their health records that are stored in data banks, inclusive of their genetic profiles, accessed by duly qualified researchers who are engaged in developing the effectiveness profiles of the interventions that could be used to deal with genetic factors that are involved in the relevant health conditions.

8.10 Addiction[31]

One of the points that were raised in the earlier part of this chapter was that the strength of the right to health care depends in part on whether the need for such care

[30] Hijmans (2018, Forthcoming).

[31] Berrettini (2017), Wang et al. (2019), Al-Eitan et al. (2021).

is auto- or hetero-induced. It was suggested that if the condition in question was auto-induced, then an individual's right to the care that would otherwise be entailed by the individual's condition would be reduced in strength relative to the claims of other persons whose same condition was not essentially auto-induced. This seems to entail that in cases of health care conditions that are the result of addiction one is dealing with auto-induced conditions, and therefore that the strength of the corresponding right to health care would invariably be reduced in comparison the right for treatment in the case of the same conditions but that were hetero-induced.

The reason that this presents an ethical issue is not that addiction is a worldwide phenomenon,[32] but that it is a health condition and that the cost of meeting the health care needs that are caused by addiction runs into billions of dollars.[33] Therefore, if an addicted person's right to care for a health condition that was causally related to their addiction had the same strength as a person's right to care for the same condition but that was not thus causally related, then this would oblige society to treat them both in the same way. This not only means that when it came to selective allocation at the hands-on level addiction would not play a role, but it would limit the resources that a society has at its disposal for meeting its health care obligations in general. This, in turn, would have the obvious implication that the types of health-oriented interventions that the society would be able to provide for its members would be reduced, and that the right to health care of its members would be more limited in execution than it would otherwise be.

The issue of competing rights to health care in general was already dealt with in the preceding parts of this study, as was the issue of competing rights in the context of auto- versus hetero-induced conditions. It was then suggested that if a condition was auto-induced, then the strength of the corresponding right to treatment was reduced and in some cases may even lapse. However, a question that was not dealt with was whether this also applies in the case of health conditions that derive from addiction.

At first glance, that does indeed seem to be the case. After all, the fact seems to be that becoming addicted is ultimately a matter of choice and that therefore addiction itself is an auto-induced condition: more correctly, perhaps, that addiction that is not transmitted from mother to child during gestation or that is not foisted on someone in some fashion is an auto-induced condition and that consequently health conditions that are the result of addiction are also essentially auto-induced. Arguably, therefore, it seems to follow that what was suggested above about the reduced strength of the right to health care for conditions that are auto-induced also applies in the case of health conditions that result from addiction.

Before proceeding any further, it may be useful to clarify the notion of addiction itself because the term covers a wide range of factors. Addiction, then, as it will be understood in what follows, is the compulsive, chronic, physiological or psychological need for a substance, behaviour or activity that has harmful physical, psychological or social effects. Moreover, typically, withdrawal or abstinence from

[32] United Nations, World *Drug Report 2020*, https://wdr.unodc.org/wdr2020/.

[33] National Institute on Drug Abuse (NIH), *Costs of Substance Abuse*, https://archives.drugabuse.gov/trends-statistics/costs-substance-abuse.

using the substance or abstaining from the behaviour or activity in question may result in conditions that require health care intervention. Therefore addiction is not merely a matter of substance abuse but includes things like compulsive exercise,[34] eating[35] and even working.[36]

With this being understood, therefore, and returning to the matter at issue, the question is whether health conditions that are the result of addiction are essentially auto-induced and therefore have reduced strength relative to the right to otherwise appropriate health care interventions.

As a matter of pure logic, the answer depends on whether the addiction in question that leads to such health conditions is entirely within the control of the individual who becomes addicted. Unfortunately, however, the data on this issue are not very clear. Thus, it appears that how an individual reacts to a given substance may depend not only on the individual's developmental stage[37] but is also functionally related to the individual's sex[38] and to their genetic makeup irrespective of age,[39] and consequently that this may vary for common addiction-associated substances like opiates and cocaine. Even how something as common as alcohol is metabolized may vary from genetic population to genetic population,[40] and consequently may give rise to different addiction profiles. Similar reasoning applies in the case of addiction that is not substance-related but centres in behaviour such as exercise.[41] Furthermore, it is arguable that the situational or social context in which an individual finds themselves and over which they have little or no control may lead to becoming addicted.[42] Consequently it is not clear that the criterion of autonomous choice that must be met in order for a condition to count as auto-induced is clearly and reliably established in the case of addiction itself.

Arguably, therefore, since it cannot be reliably established that a particular addiction is independent of such factors, it would be ethically inappropriate to scale back the strength of the right to a particular health care intervention simply because that condition is the result of an addiction or of addictive behaviour. Furthermore, the average health professional is not really qualified to judge how the addiction came about in the first place. As well, given the current state of the science, even mental health professionals who have expertise in the mental health implications of social conditions can do no more than make an educated guess in this regard. However, to base the reduction in the reduced strength of the right to health care on a guess—even on an educated opinion—is ethically questionable.

---

[34] Weinstein and Weinstein (2014).

[35] Lennerz and Lennerz (2018).

[36] Grant and Chamberlain (2016), Zou et al. (2017).

[37] Fernandez et al. (2011).

[38] Nasser and Afify (2019).

[39] Kreek et al. (2012).

[40] Ferraguti et al. (2015), Farris et al. (2010).

[41] Cf. Weinstein et all, note 34 *supra*.

[42] Eitan et al. (2017).

8.11 Autonomy and "All Other Things Being Equal"

On several occasion in the preceding discussion it was suggested that the autonomous exercise of an individual's right to health care was subject to the condition of "all other things being equal" without this qualifier being properly spelled out. In particular, this locution was used in the discussion of informed consent. To leave matter there, however, would be to leave a gap in the analysis. Before concluding the analysis, therefore, it is important to clarify the reason for this modification and in what sense and under what circumstances there is a legitimate limitation to an individual patient's autonomy in health care decision-making. It is not merely that there are certain circumstances under which an individual has only a limited right to autonomy in matters of health care, but that there are certain circumstances—for instance, when the Principle of Nonmaleficence is engaged—under which an individual's right to health care in effect turns into a duty and that, correspondingly, under such circumstances the duty that society has to provide health care services for its members also becomes duty on part of society to give formal recognition of the fact that the relevant members have a duty to avail themselves of the relevant services.

To fill in the gap, then, there are two reasons for adding this qualifier and for making these claims: first, that if an individual who suffers from certain health conditions but these are not addressed, then they pose a risk to the health of others; and second, that the right to health care itself is functionally dependent on society's ability to provide that care. The two are distinct but connected, and whereas the first is ultimately grounded in the Principles of Autonomy and Nonmaleficence, the second is grounded in the Principles of Impossibility and Nonmaleficence.

Thus, it will be recalled that the Principle of Autonomy is to the effect that everyone has the right to self-determination but that—because the Principle of Equality entails all individuals have that right—this means that an individual's right to self-determination is subject to the equal and competing rights of others. Someone's decision not to avail themselves of certain health services like immunization for contagious diseases means that they may become carriers of that particular condition and may infect others. The individual may, of course, take steps to minimize such an occurrence by engaging social distancing: i.e., by staying at least two arm's length away from other persons,[43] by wearing a mask, etc. However, as the recent example of the COVID-19 virus pandemic has shown, doing so does not guarantee that this will prevent such infections from being passed on unless the infected individual removes themselves entirely from having any contact, directly or indirectly, with other persons. If the individual does not do so and infects other persons, then the other person is harmed and the Principle of Nonmaleficence is violated. Ethically, that is unacceptable. That is why an individual's autonomy to exercise their right to health care is subject to "all other things being equal." In other words, in this sort of case, unless the individual avoids all social contact entirely, the right to health care effectively turns into a duty.

[43] Centers for Disease Control and Prevention, "Social Distancing," https://www.cdc.gov/coronavirus/2019-ncov/prevent-getting-sick/social-distancing.html.

It is important to note that this holds not only with respect to the exercise of autonomy on one's own behalf. It also applies in the case of substitute decision-making and the values that should be used by substitute decision-makers. As was indicated before, substitute decision-makers should either use the competently expressed values of the incompetent person or, absent such values, the values that may reasonably presumed the incompetent person would use if they were able to make the decision on their own behalf. Since the rights of the incompetent person are no different from the rights of competent persons, the same limiting conditions that were outlined a moment ago also apply in their case. It follows, therefore, that substitute decision-makers should observe the same rules when making their decisions.

However, not all autonomous refusals of the right to health care involve communicable conditions. For instance, someone may refuse to accept a particular intervention because it goes against their values, disagrees with how it makes them feel, etc. It is at this juncture that the Principles of Equality and Impossibility may sometimes be engaged. If the refusal to accept such interventions increases the need for further and resource-intensive health care interventions down-stream and the individual would then accept these interventions, then by exercising their initial right of refusal the individual would be increasing the drain on the resources that society has available to satisfy its members' right to health care. In other words, it would make it impossible for society to fulfil its obligation to provide health care as efficiently and as effectively as possible. By doing that, it would deprive other members of society of interventions that are their right and that would otherwise be available to them. Consequently such an exercise of autonomy on part of the initial person would violate the Principles of Impossibility and Nonmaleficence. Therefore, in such cases, the right to health care turns into a duty to accept the relevant interventions.

8.12 Conclusion

The argument in this chapter has focused on the implications that the right to health care has by focusing on the individuals to whom this right belongs. The main thrust of the argument has been that while the right extends to a great array of health care services, it is neither absolute in the sense of being a right to all health care modalities that in principle are available, nor is it always a right in the sense of being something that the individual may exercise at their own discretion.

As to the first, it was suggested that, as is the case with all rights, an individual's right to health is conditioned in actual practice by the Principle of Impossibility. Consequently, it does not follow that simply because a particular modality is available in one society that therefore it is obligatory on every other society to provide the same modality as well. The individual society's resources are here determining.

As to the second, it was suggested that because health care be delivered only in a social context and that therefore the individuals who have a right to such care have to exist in a social setting, their right is not only limited relative to the competing rights of other persons, but also may in fact turn into a duty.

The upshot of all this is that an individual's right to health care does not exist in a vacuum. It is a right only in a social context. The ultimate question, therefore, is whether the individual wishes to exist in a social context or in pure and independent isolation.

References

Ad Hoc Committee of the Harvard Medical School to Examine the Definition of Death. 1968. A Definition of Irreversible Coma. *Journal of the American Medical Association* 205 (6): 85–88.

Al-Ahsan, A. 2009. Law, religion and human dignity in the Muslim world today: An examination of OIC's Cairo declaration of human rights. *Journal of Law and Religion* 24 (2): 569–597.

Al-Eitan, L.N., D.M. Rababa'h, and M.A. Alghamdi. 2021. Genetic susceptibility of opioid receptor genes polymorphism to drug addiction: A candidate-gene association study. *BMC Psychiatry* 21 (1): 5.

Berrettini, W. 2017. A brief review of the genetics and pharmacogenetics of opioid use disorders. *Dialogues in Clinical Neuroscience* 19 (3): 229–236.

Chen, L.A., and T.N. Fawcett. 2016. Using data mining strategies in clinical decision making: A literature review. *Computers, Informatics, Nursing* 34 (10): 448–454.

Cohena, J., T. Ashkenazib, E. Katvancand, and P. Singer. 2012. Brain death determination in Israel: The first two years experience following changes to the brain death law—Opportunities and challenges. *American Journal of Transplantation* 12: 2514–2518.

Delbaere, M., and M.C. Smith. 2006. Health care knowledge and consumer learning: The case of direct-to-consumer advertising. *Health Marketing Quarterly* 23 (3): 9–29.

Di Rienzo, A. 2006. Population genetics models of common diseases. *Current Opinion in Genetics & Development* 16 (6): 630–636.

Drake, M., A. Bernard, and E. Hessel. 2017. Brain Death. *Surgical Clinics of North America* 97 (6): 1255–1273.

Dussault, G., and M.C. Franceschini. 2006. Not enough there, too many here: Understanding geographical imbalances in the distribution of the health workforce. *Human Resources for Health* 4 (12): 4–12.

Eitan, S., M.A. Emery, M.L.S. Bates, and C. Horrax. 2017. Opioid addiction: Who are your real friends? *Neuroscience and Biobehavioral Reviews* 83: 697–712.

Farris, S.P., A.R. Wolen, and M.F. Miles. 2010. Using expression genetics to study the neurobiology of ethanol and alcoholism. *International Review of Neurobiology* 91: 95–128.

Fernandez, E., R. Perez, A. Hernandez, P. Tejada, M. Arteta, and J.T. Ramos. 2011. Factors and mechanisms for pharmacokinetic differences between pediatric population and adults. *Pharmaceutics.* 3 (1): 53–72.

Ferraguti, G., E. Pascale, and M. Lucarelli. 2015. Alcohol addiction: A molecular biology perspective. *Current Medicinal Chemistry* 22 (6): 670–684.

Grant, J.E., and S.R. Chamberlain. 2016. Expanding the definition of addiction: DSM-5 vs. ICD-11. *CNS Spectr* 21 (4): 300–303.

Hijmans, Hielke, and Raab, Charles D. 2018. Ethical dimensions of the GDPR (July 30, 2018). In Mark Cole and Franziska Boehm (eds.), *Commentary on the General Data Protection Regulation*, Cheltenham: Edward Elgar. Available at SSRN: https://ssrn.com/abstract=3222677.

Kreek, M.J., O. Levran, B. Reed, S.D. Schlussman, Y. Zhou, and E.R. Butelman. 2012. Opiate addiction and cocaine addiction: Underlying molecular neurobiology and genetics. *The Journal of Clinical Investigation* 122 (10): 3387–3393.

Lennerz, B., and J.K. Lennerz. 2018. Food addiction, high-glycemic-index carbohydrates, and obesity. *Clinical Chemistry* 64 (1): 64–71.

McNary, A. 2014. Consent to treatment of minors. *Innovations in Clinical Neuroscience* 11 (3–4): 43–45 for US variations.

Mehrotra, S, and S. Kapoor. 2009. Gender discrimination in Asia: A regional perspective. *Global Social Policy* 9 (Supp): 197–205.

Nasser, S.A., and E.A. Afify. 2019. Sex differences in pain and opioid mediated antinociception: Modulatory role of gonadal hormones. *Life Sciences* 237: 116926. https://doi.org/10.1016/j.lfs.2019.116926. Epub 2019 Oct 12. PMID: 31614148.

Nicanor Pier Giorgio Austriaco, O.P. 2016. A philosophical assessment of TK's autopsy report: Implications for the debate over the brain death criteria. *Linacre Quarterly* 83 (2): 192–202.

Ngiam, K.Y., and I.W. Khor. 2019. Big data and machine learning algorithms for health-care delivery. *Lancet Oncology* 20 (5): e262–e273. https://doi.org/10.1016/S1470-2045(19)30149-4. Erratum in: *Lancet Oncology* 20 (6): 293.

Panneerchelvam, S., and M.N. Norazmi. 2003. Forensic DNA profiling and database. *Malaysian Journal of Medical Sciences* 10 (2): 20–26.

Papanicolas, I., A. Marino, L. Lorenzoni, et al. 2020. Comparison of health care spending by age in 8 high-income countries. *JAMA Network Open* 3 (8): e2014688. https://doi.org/10.1001/jamanetworkopen.2020.14688.

Sheiner, L. 2014. Why the geographic variation in health care spending cannot tell us much about the efficiency or quality of our health care system. *Brookings Papers on Economic Activity*: 1–72. https://www.brookings.edu/wp-content/uploads/2016/07/Fall2014BPEA_Sheiner.pdf. Accessed 06 Feb 2021

Suvarna, V. 2010. Phase IV of Drug Development. *Perspectives in Clinical Research* 1 (2): 57–60.

Vase, L., and K. Wartolowska. 2019. Pain, placebo, and test of treatment efficacy: A narrative review. *British Journal of Anaesthesia* 123 (2): e254–e262. https://doi.org/10.1016/j.bja.2019.01.040 Epub 2019 Mar 20.

Wang, S.C., Y.C. Chen, C.H. Lee, and C.M. Cheng. 2019. Opioid addiction, genetic susceptibility, and medical treatments: A review. *International Journal of Molecular Sciences* 20 (17): 4294.

Weinstein, A., and Y. Weinstein. 2014. Exercise addiction-diagnosis, bio-psychological mechanisms and treatment issues. *Current Pharmaceutical Design* 20 (25): 4062–4069.

White House Commission on Complementary and Alternative Medicine Policy. 2002. *Final Report*; https://web.archive.org/web/20130215040532, http://www.whccamp.hhs.gov/finalreport.html.

World Health Organization. 2020. *A Practical Guide for Health Researchers*. https://www.who.int/publications/i/item/a-practical-guide-for-health-researchers

Zou, Z., H. Wang, F. d'Oleire Uquillas, X. Wang, J. Ding, and H. Chen. 2017. Definition of substance and non-substance addiction. *Advances in Experimental Medicine and Biology* 1010: 21–41.

Appendix
Health Care Corporations

General Considerations

The chapters of this study dealt with the ethical issues that surround the right to health care—and, indeed, whether there is such a right in the first place—from the perspective of society, members of society, professionals and health care institutions.

However, these are not the only players in the field of health care. There also are the corporations that supply many of the tools and devices that are used in the provision of health care and without which the provision of modern health care would be unimaginable. These tools and devices range in nature from diagnostic to interventional items, and may be material, electronic or chemical in nature. While the research for of these may involve health care institutions as well as patients and health care professionals, their actual development and manufacture falls outside of the realm of hands-on health care and is logically distinct from the framework within which that care is delivered.

At the same time—not to go back on what was just said—in an obvious sense these distinct corporate entities do operate within the general field of health care: They not only provide the tools and devices that are used in the delivery of hands-on health care but also in many cases—there are a few exceptions, particularly in the diagnostic field—they would not even exist if it were not for the field of hands-on health care itself.

This notwithstanding, however, the two—i.e., those who actually deliver health care and the corporate bodies that provide the tools and devices that are used in doing so—are importantly different. What distinguishes the corporate bodies that provide the tools and devices is that, their operating in the general field of health care notwithstanding, what they develop and the fact that what they produce is used in the delivery of health care is really incidental to the reason for their development and production. For, the mandate of these corporate bodies is not to improve the health of those who receive their products and who are treated with them. While that may be the effect that the of use their products may have, and while the improvement of

E.-H. W. Kluge, *The Right to Health Care: Ethical Considerations*, The International Library of Bioethics 92, https://doi.org/10.1007/978-3-030-93838-3

patient health may be the aim of those who use or prescribe them, the mandate and aim of these corporate bodies—indeed, their essential role—is to provide their stockholders with a return on their investments. In other words, their mandate is purely commercial. And while the relationship between such corporations and their stockholders is subject to ethical constraints,[1] these constraints are inherently different from those that characterize health care institutions and health care professionals.

At the same time, however, and simply because both the development and use of their products falls within the general field of health care, the ethical issues that arise for these corporate bodies straddle the two realms and require solutions that are defensible in both realms. Moreover, these issues are functionally determined by how health care itself is understood. More specifically, when health care is construed as a commodity, the issues that arise essentially become contractual issues of implied warranty—e.g., whether the product is fit for the purpose—whereas when health care is construed as a right the ethical issues that arise involve matters that are grounded in the principles that structure the rights-approach to health care itself and contractual issues form only part of what is at stake.

A good example of this is the development and production of pharmaceuticals. In most cases the medications that are used in the delivery of health care are developed and produced by pharmaceutical companies. This means that while they must be approved as medically appropriate, effective and safe by a duly constituted licencing body of society, their development and production is not really a matter of societal health care obligation except in the sense that society has a duty to provide for relevant and appropriate research that deals with otherwise unmet health care needs. Moreover, if the medication is new, non-obvious and not something that is a purified version of something that is found in nature, the company that develops and produces it is deemed to have an intellectual property right in it and may assert this right by obtaining a patent.[2] This is outlined in the *Agreement on Trade-Related Aspects of Intellectual Property Rights*, which is a legal agreement between all the members of the World Trade Organization. In a document called "Trade-Related Aspects of Intellectual Property Rights" (TRIPS)[3] it sets out the standards that national governments may use to protect different forms of intellectual property. The purpose of TRIPS is to permit countries to protect the financial interests of their corporations when these develop new products so that, among other things, the companies can recover the costs that are involved in developing them and bringing them to market.

Applied to the present context, the reason—the practical reason—for all of this is that the costs that are involved in developing a new pharmaceutical product and bringing it to market are not small. It is estimated that the median capitalized research and development cost for a single pharmaceutical product is $985 million

---

[1] Boatright (2009, Crane et al. (2008), French (1995).

[2] World Ttrade Organization, *Patent Law*; accessed 05/12/2020 at https://www.wto.org/english/thewto_e/acc_e/cgr_e/WTACCCGR27A3_LEG_9.pdf.

[3] World Trade Organization, *Agreement on Trade-Related Aspects of Intellectual Property Rights* (TRIPS), accessed 28/11/2020 at https://www.wto.org/english/docs_e/legal_e/27-trips.pdf.

US, counting expenditures on failed trials[4]—which latter amount to approximately 54% of all development pharmaceutical research that is undertaken.[5]

Arguably, it is in the interest of a society that wishes to provide an equitable level of health care for its members to have appropriate and effective pharmaceuticals in its health product armamentarium. However, unless society itself engages in the relevant research, these will not exist unless they are researched and developed by pharmaceutical corporations. And, because the corporations have the ethical obligation to provide a return-on-investment for their stockholders, the expenses involved in the research and development of the medications must at least be recovered.

However, the ethically complicating factor is that patients are involved in the research and development process. While that process does not always involve patients—patients are generally not involved in Phase 0 trials, which are intended to determine how a pharmaceutical is metabolised in a healthy body[6]—patients are integrally involved in Phase II-III trials and sometimes in Phase I trials as well.[7] Moreover, while the health care professionals are usually reimbursed for their services by the corporations, the patients in Phase II-III trials are not; and while patients generally participate in Phase II-III trials in the hope that the they will receive some therapeutic benefit from doing so, this is not a matter of valid expectation on their part, as would be the case when they accept a prescription that is prescribed by an attending health care professional for a particular condition.

Arguably, therefore, all of this means that in a rather obvious sense pharmaceutical companies may well be obligated to the patients who volunteer to participate in their Phase II–III trials. The question therefore is how this obligation, insofar as it exists, can be discharged or whether in fact it is an obligation at all.

Societal Issues

This latter question will be addressed more directly in a moment. For now, there is another question that arises, and it concerns society itself. Does society have an ethical right to limit the intellectual property rights that a corporation has in the therapeutic health care related products that it has researched and developed?

The reason that this becomes important is that if a society adopts the position that health care is a right but it does not restrict the intellectual property rights that a corporation has in what it has researched and developed, then the corporation may essentially set the price that it charges for its products. This, however, when the price is steep, may seriously impair society's ability to fund at an appropriate level the

[4] Wouters et al. (2020).

[5] Fogel (2018).

[6] US Food and Drug Administration "The Drug Development Process," https://www.fda.gov/patients/learn-about-drug-and-device-approvals/drug-development-process.

[7] World Health Organization, *Clinical Trials*, https://www.who.int/health-topics/clinical-trials/#tab=tab_1.

health care that it is ethically obligated to provide. The controversy that surrounded the fees that Myriad Genetics initially charged for the tests that it developed for BRCA1 and BRCA2 is here a case in point,[8] as is the case of BioNtech and the price it allegedly wanted to charge the European Union for its COVID-19 vaccine.[9]

Part of the answer to the question is found in the TRIPS Agreement itself. Thus, Sects. 27 and 30–31 of that document stipulate that "legitimate interests of third parties" constitute an exception to the exclusivity of use and production that is granted under patent law, where the "legitimate interests of third parties" is not to be construed in a purely legal sense but in the sense of what is generally accepted in the social context. This, then, means that if a society construes health care as a right, then a corporation's right to determine the price for its product could legally be limited by that society if the society could show that doing so would be in the "legitimate interests of third parties." For, clearly, access to appropriate and effective pharmaceuticals would be in the interest the members of such a society; and while pricing the interventions at a level that exceeded the cost of their development might well advance the interest of the stockholders of the corporations, it would not necessarily be in the interest of the members of society who required the medication.[10]

However, this is merely a legal matter and does not touch the ethics. As to the ethics, it could be argued that legal considerations notwithstanding, the above reasoning would be ethically valid only if it included as premises the following theses: (1) that the rights of the majority trump the rights of the minority—i.e., that the right of the members of society trump the right of the investors of the pharmaceutical corporations—and (2), more specifically, that the right to affordable health care trumps the financial rights of investors.

However, in light of what was argued in Chap. 2, premise (1) would mean that it would involve a switch from a deontological to a utilitarian ethical approach. If that was accepted it would mean that the argument that health care is a right and not a commodity as this was presented in the subsequent chapters would no longer be valid. In that case, however, the issue itself would disappear as an ethically relevant issue because then there would be no right to health care in an ethically relevant sense. There would only be a right to health care in a legal sense, and the apparent conflict could be resolved by appropriate legal means.

As to premise (2), it would require showing that different types of rights have a ranking order relative, and aside from what was outlined in Chaps. 3 and 4—which only applies to the ranking of rights within health care—it is not quite obvious how such an argument would proceed in this sort of case. This, then, seems to present a conundrum.

However, in opposition to this it could be argued that this conundrum arises only if the framework for considering the issue is drawn too narrowly. The claim that society

---

[8] Eggertson (2002).

[9] *Frankfurter Allgemeine Zeitung* "Biontech wehrt sich gegen Wucherpreis-Berichte", https://www.faz.net/aktuell/wirtschaft/unternehmen/corona-impfstoff-biontech-wehrt-sich-gegen-wucherpreis-berichte-17207979.html.

[10] Lee Mendoza (2019).

would have a right—indeed, would have a duty—to place limits on the control that a pharmaceutical company has over the pricing of the products that the company patents does not depend on switching to a utilitarian ethical approach, nor do the interests—more correctly, the rights—of the stockholders have to be bow to the rights of members of society. Both conditions can be reconciled by noting that both the right to health care and the right of the stockholders both arise in a social context.

More specifically, the right of stockholders to a return on their investment is not something that obtains in a vacuum. Not even the Lockean thesis that labourers have a right to the fruits of their labour[11] alters this fact, because the labour that would be involved in this case would itself be possible only in the social framework in which corporations and investments exist and function. If it is to be ethical, however, the social framework itself would have to be governed by the fundamental ethical principles that were outlined in Chap. 2—minimally, with respect to the Principle of Impossibility, which has both ethical and legal validity.[12] This in turn means that, legal considerations aside, the rights that arise out of the social interactions that are involved in the functioning of a corporation are conditioned by the same principles as all other rights. That, however, means that the right of return-for-investment has to be reconciled with the right to equitable access to health care.

As was argued in the preceding chapters, an ethical society has a duty to provide equitable access to health care for its members. As was also suggested, this means that an ethical society has an obligation to develop appropriate, effective and safe health care modalities—and that includes developing appropriate, safe and effective medications. This, in turn, means that it must either itself engage in the research and development that are necessary to produce such modalities, or it must allow the formation and functioning of independent corporations that will do so.

However, for a society to engage in the relevant activities does not mean that costs would not be involved, nor does it mean that society could ignore these costs and simply write them off. Society, after all, only has a limited amount of resources, which in turn means that the costs would have to be recovered and could not simply be syphoned off from other areas of socially mandated services, for that would prevent society from meeting the obligations that it has in these other domains. Moreover, it would still be necessary to have administrative or corporate structures to plan and carry out the relevant research and developmental activities because, as was also pointed out before, doing so is not the same as delivering health care. In fact, logically all of this would have to precede the delivery of health care because the delivery of health care depends on having the appropriate and relevant tools. Therefore, the best that society could do if it chose to adopt this option would be to set the price structure at a cost-recovery level.

However, if a society did not wish to adopt this option, then it would have to allow the formation and operation of distinct corporate bodies that would engage in the research and manufacture of appropriate, effective and safe pharmaceuticals,

---

[11] Locke (1690).

[12] Chapter 2, p. 23.

and that these corporations would then be able to apply for and receive patents. This means that they would then be legally entitled to set the price of their products.

It is arguable, however, that on this option society could not reasonably insist that this price be set on a cost-recovery basis because this would remove the incentive for corporate bodies to invest in the research and development of the pharmaceuticals in the first place In fact, it could be argued that from a purely ethical perspective it would be problematic for a corporation to do so because, as was pointed out before, the corporation has an ethical obligation to provide its stockholders with a return on their investment, and that operating on a pure cost-recovery basis would not meet that obligation. Therefore if society were to adopt this second option, the only ethically appropriate thing for it to do would seem to be for society to set what could be considered a reasonable margin of profit in such cases.

However, that would also seem to be not without problems. For, it would seem that unless society were at the same time to set such a reasonable-margin-of-profit limit on all corporations that provided products and services, it would be violating the Principle of Equality and therefore would be unethical. Moreover, this would also be counterproductive and would have the effect of undermining the research and development of appropriate, effective and safe pharmaceuticals. Corporations would then simply direct their efforts into areas where such limits did not exist. Nor would it be unethical for corporations to engage in different and non-pharmaceutical oriented endeavours because, *qua* corporations, they do not have a duty to engage in the research and development of tools that would enable society to meet its duty to provide health care. As was said before, their ethical obligation is to provide their stockholders with a return for their investment. The question of how intellectual property rights in pharmaceuticals and related health care products should ethically be handled therefore seems to remain a conundrum. While it could be settled in practice on the basis of political, legal and even purely emotional considerations, this would not guarantee the ethical validity of the solution.

However, there is another aspect to the matter that so far has been ignored: the fact, namely, that the issue of cost control enters only in contexts where society has an obligation to provide particular kinds of services; because it is only in this sort of context that the issue of society's limited resources and the duty to provide services overlap. Therefore the argument that society would not be acting ethically unless it imposed a similar margin-of-profit limitation on all corporations within its jurisdiction is not really valid because not all corporations develop and produce products that fall under this rubric. Therefore the claim that society could not ethically impose a profit-margin control unless it did so across the board is fallacious.

As to the argument that it would be counterproductive because corporations would simply direct their investments to areas where there was no such profit margin control, that is based on the unwarranted assumption that the corporations would inevitably do so. However, as is clear from stock markets and financial investment reports, that is not the case. Nor would it be valid to argue that unless the corporations followed such a strategy, they would violate their obligation to their stockholders. That would be the case only if they acted in a way that predictably would not return a profit for their investors—and that would not here be the case. It would merely be that the return would have limits. At the same time, it would essentially be a guaranteed return, since society would be forced to purchase the products in order to meet its health service obligations.

Orphan Drugs

Closely related to the preceding issue is the question of how an ethical society should handle the sort of situation where the market for the relevant health care products was so limited that it would not be in a corporation's financial interest to venture into this area because it could never expect to meet the cost of researching and producing the product, let alone reap a profit from such an investment. With respect to pharmaceuticals, this has become known as the issue of so-called "orphan drugs." More specifically, there are health conditions whose incidence and prevalence is so low that the cost for researching and developing appropriate, safe and effective medications to deal with them would exceed by far what it would cost to do so. The general understanding is that these are conditions that affect no more than 5 or 6 people in 10,000.[13]

However, the fact that the conditions are rare does not mean that an ethical society that construes health care as a right does not have an obligation to provide appropriate, safe and effective health interventions for those who suffer from them if the conditions interfere with a person's ability to take equitable advantage of the opportunities that are otherwise available to the society's members. Nor does it mean that merely providing palliative services for such individuals would meet this obligation. The Principle of Equality entails that unless the Principle of Impossibility entered the picture—which is to say, unless it was actually impossible to deal with the condition—the duty to provide equitable access to health care services entails that the same effort should be expended to provide mitigating treatments for these conditions as is expended for the persons who suffer from the more common conditions.

Society therefore once more seems to be faced with two options: It could either structure its health care system in such a way that it included a branch whose purpose was to research and develop appropriate, safe and effective treatment for these conditions—where of course, by the Principle of Equality, the degree of appropriateness, effectiveness and safety would have to be similar to that which exists in the treatments of more common conditions—or it would have to take the steps that would be necessary to encourage free-standing corporate entities to engage in the relevant research and development without the corporations violating their obligations to their stockholders.

The first option was already discussed and requires no further comment. As to the second option, such steps could range from providing tax incentives that would

[13] Giannuzzi et al. (2017).

allow the companies to claim a significant amount of the relevant costs as tax write-offs (as is the case in The People's Republic of China[14] and some other countries) to providing exclusive marketing rights for a given period of years,[15] fast-tracking approvals of drugs intended for rare diseases,[16] and even providing special funding whose purpose would be to defray the research and development costs that were involved and which would not be recovered unless the price of the medications reflected the unsupported cost involved in bringing them to market.[17] In any case, however, if a society decided not itself to engage in the research and development itself, it could not ethically insist that the corporations not make a profit. At best, it could adopt the limitation-of-profit option that was discussed above.

Corporate Behaviour on a Global Level

Another set of ethical issues that arises for corporations that engage in the research and development of the tools that are used in the delivery of health care—and this again applies particularly, although not exclusively, to pharmaceutical corporations—is whether these corporations have an obligation to reward—and if so, how or in what manner—the subjects who are involved in the various stages of that research and development.

To put this into context, a pharmaceutical corporation will attune the research and development for its products to the population that suffer from conditions for which these will be therapeutically appropriate. In many cases, this applies globally because most human beings will be susceptible to similar health conditions. However—and this seems to potentiate an ethical problem—not all societies have promulgated the same standards that must be met in order for the relevant research and development to be carried out, nor do all societies have the same rate of incidence and prevalence of the relevant conditions and therefore could not provide a sufficiently large subject pool to meet the scientific standards that are necessary for approval of the intervention as being safe and effective.

[14] Jin and Chen (2016).

[15] Gericke et al. (2005).

[16] U.S. Dept. of HHS, FDA, CDER, and CBER, *Guidance for Industry: Expedited Programs for Serious Conditions – Drugs and Biologics*; available at https://www.fda.gov/regulatory-information/search-fda-guidance-documents/expedited-programs-serious-conditions-drugs-and-biologics.

[17] US Food And Drug Administration, *Orphan Products Grants Program*; accessed 08/12/2020 at https://www.fda.gov/industry/developing-products-rare-diseases-conditions/orphan-products-grants-program.

Venue Shopping

The corporations therefore tend to do what is sometimes called "venue shopping": i.e., to conduct their research and development in jurisdictions where the relevant regulations are less strict than in others, or in societies where the incidence and prevalence of the condition are sufficiently high to make it easier to recruit a sufficiently large subject pool to meet the scientific validity condition that would usually be necessary to obtain approval of the tool, substance or intervention.

The central ethical issue that is presented by this is not the appropriateness of choosing a society where incidence and prevalence of the condition are sufficiently high to allow the recruitment of a sufficiently large subject pool to meet the relevant standards of scientific validity. That is merely good science and good practice. The ethical issue is that in many cases the society that is selected also suffers from a comparatively low standard of living, and that therefore enrolling the relevant subjects can more easily be achieved by offering otherwise inappropriate levels or types of inducement for consenting to participate in the research protocols. If the analysis of research in the preceding chapters is valid, then doing so would be unethical. However, while unethical corporate behaviour can be controlled within a particular jurisdiction, it is not clear how—indeed, whether—this can be done on a global level. Therefore this raises the question whether an ethical society is obliged to ensure that health care corporations that are incorporated in their jurisdiction act ethically not only within their own jurisdiction but also in whatever jurisdiction they operate.

Business Versus Societal Ethics

By the same token, but this time from the side of the corporations, there is the question whether the obligation that the corporation has towards its stockholders takes priority over the rights that the members of the society in which it conducts its research and development. This latter issue particularly arises in situations where the society in which it does its research and development has comparatively lax legal requirements for engaging in such research. Should the corporation follow the legal requirements as these are stated in the relevant jurisdiction, or should the corporation structure it protocols in keeping with the fundamental principles of ethics that were outlined in Chap. 2?

Not surprisingly, given what has been argued in the preceding chapters, the answer is that a corporation is ethically obligated to structure its protocols in keeping with the fundamental principles of ethics. If the society in which it operates does not to structure its rules and regulations in accordance with these principles, that would be an ethical failure on part of the society, but that would not exonerate the corporation itself. Therefore while it would be a legal excuse for a corporation not to adhere to these principles, it would not be an ethical excuse. Law is not ethics—as was driven home in the judgement of those who participated in the extermination of the inmates in Nazi concentration camps when their claim that they were merely following then legal orders was not accepted, or when those who instituted and participated in the genocide in Srebrenica were found guilty of crimes against humanity. Consequently it would not serve as an acceptable ethical excuse for the corporations to follow merely legal requirements in their research and development protocols.

Another question is whether a corporation owes an ethical debt to the individuals whom it enrols in its research and development projects and, more generally, whether it owes an ethical debt to the society of which these individuals are members. The reason for the first part of this question is that participation in Phase I-III trials is on a voluntary basis, and while some inducement may be offered to encourage participation, it is generally agreed that it should not overwhelm the autonomy of the individuals and thereby constitute what is referred to as undue enticement.[18] An element of beneficence on part of the volunteers, therefore, remains and the question is whether this would in some way ethically entitle the volunteers to some recognition by the corporation. In particular, the question becomes whether the members of the society to which the volunteers belong are entitled to receiving the medication that is developed at a lower price, at cost or even below cost. Once more, the reason for this question is that, as was indicated above, the societies of which the volunteers are members are often societies that are socio-economically disadvantaged and could not afford the costs of purchasing the developed products even if the cost were set at what would otherwise be considered a fair and reasonable rate.

Some of the ethical considerations that would deal with this question draw on the Principle of Autonomy, and involve considerations about how autonomy ethically functions in actual situations. More specifically, if the inducement that was offered to potential participants was adjusted to the socio-economic and other relevant conditions of the potential subjects, then it seems reasonable to suggest that the inducement would not constitute what is called undue enticement, and participation would be on a voluntary and beneficent basis, the fact of inducement itself notwithstanding.

However, there is more to the mater. Ethically speaking, subject participation in research is on a voluntary basis. Logically, therefore, the fact that participation is on a voluntary basis entails that the volunteers—their hopes and expectations aside—would be acting in keeping with the Principle of Beneficence. Beneficence, however, is not a trade item but an ethical principle that applies to social interactions. As was suggested in Chap. 2 it characterizes the structure of obligations that individuals have towards other individuals. Arguably, therefore, while it would be appropriate to respect individuals who are beneficent in their actions, that does not entitle them to special considerations. They are merely acting in an ethical praiseworthy fashion. In other words, it could be argued that corporations—and, again, particularly pharmaceutical corporations—would not have an ethical duty to offer the relevant products at a reduced rate to the individuals who participate as volunteers in their research and development.

---

[18] Cf. Tri-Council Guidelines, *Policy Statement: Ethical Conduct for Research Involving Humans – TCPS 2 (2018)*, https://ethics.gc.ca/eng/policy-politique_tcps2-eptc2_2018.html, especially Article 3 et pass (Art. 3.2 and 11.2).

Once again, however, and irrespective of where the corporations are incorporated, to leave matters there would be to overlook the fact that the corporations would be conducting their research and development in a given society. This means that the particular society's ethical framework has a role to play in how the corporation carries out its business. Therefore if the society construes health care a as right—and if the argument in the preceding chapters is valid, then it should do so—then the society has an obligation to structure the juridical framework within which the corporation operates in a way that not only respects the ethics of research as this was outlined in the preceding chapters, but also to ensure that the cost of the relevant health care products would not exceed society's ability to provide the services that would ultimately be developed and that would rely on them. This, in turn, would mean that it could ethically impose a legal duty on the corporation to provide the product that it develops in its jurisdiction at a price that was consistent with the society's ability to pay. This, of course, if adopted, could very well mean that the prices that such a corporation charges for its product could vary from society to society without violating the Principle of Equality. However, in order to have consistency in this regard, a global authority would have to be established whose role would be to determine what would be ethically appropriate for a given society, perhaps along the lines of the European Union's adjudication mechanism for its General Data Protection Regulation.

It is possible, therefore, that unless the whole field of health care was restructured in a global sense and became a global matter, and unless the right to health care was considered to be something that consistently belonged to all persons irrespective of their particular societal affiliation, the issue of how the products and devices that are used in health care in general and of pharmaceuticals in particular should be controlled and priced will remain a conundrum.

References

Boatright, J.R. 2009. *Ethics and the Conduct of Business*, 6th ed. Upper Saddle River, NJ: Pearson Prentice Hall.

Crane, A., D. Matten, and J. Moon. 2008. *Corporations and Citizenship*. Cambridge University Press.

Eggertson, L. 2002. Ontario defies US firm's genetic patent, continues cancer screening. *Canadian Medical Association Journal* 166 (4): 494.

French, P.A. 1995. *Corporate Ethics*. Harcourt Brace.

Fogel, D.B. 2018. Factors associated with clinical trials that fail and opportunities for improving the likelihood of success: A review. *Contemporary Clinical Trials Communications* 11: 156–164.

Gericke, C.A., A. Riesberg, and R. Busse. 2005. Ethical issues in funding orphan drug research and development. *Journal of Medical Ethics* 31:164–168.

Giannuzzi, V., R. Conte, A. Landi, et al. 2017. Orphan medicinal products in Europe and United States to cover needs of patients with rare diseases: An increased common effort is to be foreseen. *Orphanet Journal of Rare Diseases* 12: 64.

Jin, X., and L. Chen. 2016. Orphan drug development in China—Turning challenges into opportunities. *Intractable Rare Dis Res.* 5 (4): 308–313.

Lee, Mendoza R. 2019. Incentives and disincentives to drug innovation: Evidence from recent literature. *Journal of Medical Economics* 22 (8): 713–721.
Locke, J. 1690. *Second Treatise of Government*, II, Section 6.
Wouters, O.J., M. McKee, and J. Luyten. 2020. Estimated Research and Development Investment Needed to Bring a New Medicine to Market, 2009–2018. *JAMA* 323 (9): 844–853. https://doi.org/10.1001/jama.2020.1166

Index

E.-H. W. Kluge, *The Right to Health Care: Ethical Considerations*, The International Library of Bioethics 92, https://doi.org/10.1007/978-3-030-93838-3

The manufacturer's authorised representative in the EU is Springer Nature Customer Service Centre GmbH, Europaplatz 3, 69115 Heidelberg, Germany. If you have any concerns regarding our products, please contact ProductSafety@springernature.com

Printed and bound by CPI Group (UK) Ltd, Croydon, CR0 4YY
30/09/2026
02225690-0009